INTERMEDIATE MATHEMATICAL
ANALYSIS

Intermediate Mathematical Analysis

Hugh Thurston
Formerly University of British Columbia

CLARENDON PRESS · OXFORD
1988

Oxford University Press, Walton Street, Oxford OX2 6DP
Oxford New York Toronto
Delhi Bombay Calcutta Madras Karachi
Petaling Jaya Singapore Hong Kong Tokyo
Nairobi Dar es Salaam Cape Town
Melbourne Auckland
and associated companies in
Berlin Ibadan

Oxford is a trade mark of Oxford University Press

Published in the United States
by Oxford University Press, New York

British Library Cataloguing in Publication Data
Thurston, Hugh
Intermediate mathematical analysis.
1. Functions of several variables.
Differentiation
I. Title
515.8'4
ISBN 0-19-853291-1
ISBN 0-19-853292-X Pbk

Library of Congress Cataloging in Publication Data
Thurston, H. A. (Hugh Ansfrid)
Intermediate mathematical analysis / Hugh Thurston.
Includes indexes.
1. Mathematical analysis. I. Title.
QA300.T48   1988   515—dc19   88-6973
ISBN 0-19-853291-1
ISBN 0-19-853292-X (pbk.)

Typeset and printed by The Universities Press (Belfast) Ltd

Preface

Intermediate analysis goes beyond the functions of one variable that are studied in elementary analysis and deals with functions of several variables. It stops short of functions on abstract spaces, whose study is classed as advanced analysis.

The chief ways in which intermediate analysis is harder than elementary analysis are due to the extra geometrical richness of higher dimensions. In one dimension we can, with no serious loss of generality, confine our attention to functions whose domains are intervals: we integrate only over intervals, and such theorems as the intermediate-value theorem and the mean-value theorem are stated for intervals. In higher dimensions we cannot do this, and consequently we need various techniques from topology. For instance, we need to be careful about continuity at the boundary points of the domain of a function, and the very word 'boundary' is a topological term.

A glance at the table of contents shows the precise ground covered by this text. I should like to comment on two points.

There are two common notations for partial derivatives. One notation, D_1f, D_2f, etc., uses numerical suffixes; the other, $\partial u/\partial x$ or u_x, does not. The D notation is usually preferred in analysis because it has a logically sound definition. To define $\partial u/\partial x$ seems to be more difficult, and many otherwise respectable texts give an unsound definition. However, a sound definition is possible in two steps, starting with a definition of D_1f. (Readers familiar with manifolds will recognize the definition we give as equivalent to defining $\partial u/\partial x$ on a given manifold.)

The 'change-of-variable' theorem for integrals is notoriously difficult and its treatment in standard texts is not always satisfactory. The traditional formula for this theorem (in two dimensions) is

$$\int_{\mathbb{A}} f(x, y) \, dx \, dy = \int_{\mathbb{B}} g(u, v) \frac{\partial(x, y)}{\partial(u, v)} \, du \, dv$$

where $f(x, y) = g(u, v)$ and the transformation from (x, y) to (u, v) transforms $\mathbb{A}$ into $\mathbb{B}$. There are two difficulties. The first is that a perfectly respectable transformation can, from a well-behaved f, yield an unbounded and therefore non-integrable g, and can transform a bounded $\mathbb{A}$ into an unbounded $\mathbb{B}$. This difficulty can be overcome by using improper integrals. (The alternative of insisting on boundedness in the statement of the theorem is too restrictive.) The more serious difficulty is that most extant proofs require the transformation to be well behaved not just on $\mathbb{A}$ but also outside $\mathbb{A}$—usually on an open set containing the closure of $\mathbb{A}$. Our proof overcomes both difficulties.

Prerequisites for this text are elementary analysis (of course) and a little linear algebra: to be precise, the reader will need to know how to multiply matrices and to know that a linear transformation has an inverse if and only if the determinant of its matrix is non-zero.

The exercises are routine questions designed to test whether the reader has understood the preceding text. The problems call for more ingenuity, and some are quite hard.

White Rock, B.C.   H. T.
1987

Contents

Notation

We use the following notation from elementary analysis.

If $\mathbb{S}$ is a set of real numbers:

$\sup \mathbb{S}$ is the supremum (the least upper bound) of $\mathbb{S}$;

$\inf \mathbb{S}$ is the infimum (the greatest lower bound) of $\mathbb{S}$;

$\sup_{\mathbb{S}} f$ is the supremum of f on $\mathbb{S}$, i.e. the supremum of $\{f(x) : x \in \mathbb{S}\}$, and similarly for $\inf_{\mathbb{S}} f$.

If x_1, x_2, etc. are real numbers, $\min(x_1, x_2, \ldots)$ is the least of them and $\max(x_1, x_2, \ldots)$ is the greatest of them.

1
Functions of several variables

We assume that the reader is familiar with the general concept of function. If f is a function, the set of all entities x for which $f(x)$ exists is the **domain** of f; each such x is an **argument** (or input) of f, and $f(x)$ is the corresponding **value** (or output).

Functions encountered in elementary calculus have real numbers for arguments, but functions whose arguments are pairs of real numbers occur quite naturally. For instance, let $v(x, y)$ cm³ be, for each positive x and y, the volume of a cone whose height is x cm and whose base radius is y cm. The arguments of v are number pairs. Note that $(1, 2)$ is not the same as $(2, 1)$, and indeed $v(1, 2) = 4\pi/3$ while $v(2, 1) = 2\pi/3$. To emphasize this we call these pairs **ordered pairs**. Ordered pairs (x, y) and (x^*, y^*) are equal if and only if $x = x^*$ and $y = y^*$.

Functions whose arguments are ordered trios occur equally naturally.

Notation   We let $\mathbb{R}$ denote the set of all real numbers, $\mathbb{R}^2$ the set of all ordered pairs of real numbers, $\mathbb{R}^3$ the set of all ordered trios of real numbers, and so on.

The arguments of the function v above belong to $\mathbb{R}^2$, and its values belong to $\mathbb{R}$. If the arguments of a function belong to $\mathbb{A}$ and its values to $\mathbb{B}$, we say that it is a function **in** $\mathbb{A}$ **into** $\mathbb{B}$. Thus v is a function in $\mathbb{R}^2$ into $\mathbb{R}$. (Some authors use the phrase 'from $\mathbb{R}^2$ to $\mathbb{R}$' instead, but other authors use this phrase only when the domain of the function is the whole of $\mathbb{R}^2$.)

We shall be studying functions in $\mathbb{R}^n$ into $\mathbb{R}$ and, more generally, functions in $\mathbb{R}^n$ into $\mathbb{R}^m$.

Because we are studying only functions of this type (and not, say, functions whose arguments are complex numbers) we may sometimes say '. . . for every x . . .' instead of '. . . for every real number x . . .', where the context makes it clear that x denotes a real number.

Examples 1. A cartesian coordinate system is set up in three-dimensional space. For every x, y, and z, let $d(x, y, z)$ be the distance between the points $(0, 0, 0)$ and (x, y, z). Then d is a function in $\mathbb{R}^3$ into $\mathbb{R}$.

2. $\mathcal{S}$ is the hemisphere $x^2 + y^2 + z^2 = 1$, $z \geq 0$. Let $q(u, v)$ be the coordinates of the point in which the line $x = u$, $y = v$ cuts $\mathcal{S}$, for every u and v for which it does cut $\mathcal{S}$. Then q is a function in $\mathbb{R}^2$ into $\mathbb{R}^3$.

We remind the reader of some familiar concepts and introduce some notation. The domain of f is denoted by $\operatorname{dom} f$. The **range** of f is the set of all values of f; we denote it by $\operatorname{rng} f$.

If $\mathbb{A}$ is contained in the domain of f, the **restriction** $f_\mathbb{A}$ of f to $\mathbb{A}$ is the function with domain $\mathbb{A}$ for which $f_\mathbb{A}(x) = f(x)$ whenever $x \in \mathbb{A}$.

If $\mathbb{A}$ is a set, the **image** $f(\mathbb{A})$ **of** $\mathbb{A}$ **under** f is the set of all $f(x)$ for $x \in \mathbb{A} \cap \operatorname{dom} f$. (Note. If we were dealing with general functions this notation would be ambiguous, and some abstract algebra texts use $f_*\mathbb{A}$ instead. For example, if 1, 2 and $\{1, 2\}$ all belong to $\operatorname{dom} f$, then $f(\{1, 2\})$ as defined here would mean $\{f(1), f(2)\}$ as well as the value of f at $\{1, 2\}$. For functions in $\mathbb{R}^n$ into $\mathbb{R}^m$, however, there is no ambiguity.)

f is **one-to-one** if $f(x)$ cannot equal $f(y)$ unless $x = y$.

If f is one-to-one the function $f^\smile$ for which $f^\smile(x) = y$ if and only if $f(y) = x$ is the **inverse** of f. (Note. We avoid the notation f^{-1} because it can be confused with the reciprocal of f.)

If $\mathbb{B}$ is a set, the **inverse image** of $\mathbb{B}$ under f is the set of all x for which $f(x) \in \mathbb{B}$. We denote it by $f^\smile \mathbb{B}$. If f has an inverse $f^\smile$, then $f^\smile \mathbb{B} = f^\smile(\mathbb{B})$.

The **composite** $g \circ f$ of f with g is the function for which $g \circ f(x) = g(f(x))$ for every x for which $g(f(x))$ exists. (Note. If f is in $\mathbb{R}^n$ into $\mathbb{R}^m$ and g is in $\mathbb{R}^q$ into $\mathbb{R}^p$, $g \circ f$ will be null unless $m = q$, in which case $g \circ f$ will be in $\mathbb{R}^n$ into $\mathbb{R}^p$.)

If f is in $\mathbb{R}^n$ into $\mathbb{R}^m$ and $\operatorname{dom} f$ is the whole of $\mathbb{R}^n$ we say that f is **on** $\mathbb{R}^n$; if $\operatorname{rng} f$ is the whole of $\mathbb{R}^m$ we say that f is **onto** $\mathbb{R}^m$. In general, we say that f is **on** $\operatorname{dom} f$ **onto** $\operatorname{rng} f$.

f is an **identity function** if $f(x) = x$ for every x in the domain of f.

f is **constant** if $f(x)$ is the same for every x in the domain of f.

Exercises
1. $\operatorname{dom} f = \mathbb{R}^3$, $f(x, y, z) = (x + y, y + z, z + x)$. Find rng f. Does f have an inverse? If so, specify it. (One way to specify a function is to give its domain and a formula for the general value; that is how we specified f.)
2. $\operatorname{dom} g = \mathbb{R}^3$, $g(x, y, z) = (x - y, y - z, z - x)$. Find rng g. Does g have an inverse? If so, specify it.
3. Let $\mathbb{A} = \{(x, y, z): x > 0, y > 0, \text{ and } z > 0\}$ and f and g be as in Exercises 1 and 2. Find $f(\mathbb{A})$, $g(\mathbb{A})$, and the inverse images of $\mathbb{A}$ under f and g. Specify $f \circ g$, $g \circ f$, $f \circ f$, and $g \circ g$.
4. $\operatorname{dom} f = \mathbb{R}^2$, $f(x, y) = 2x + 3y$, $\operatorname{dom} g = \mathbb{R}$, $g(x) = (x, -x)$. Specify $g \circ f$ and $f \circ g$. What can you say about $f \circ f$ and $g \circ g$?

We now introduce some concepts special to functions in $\mathbb{R}^n$. If f and g are in $\mathbb{R}^n$ into $\mathbb{R}$, we define (f, g) to be the function for which

$$(f, g)(x) = (f(x), g(x))$$

for every x for which the right-hand side exists. Then (f, g) is a function in $\mathbb{R}^n$ into $\mathbb{R}^2$. A similar definition applies for three or more functions. If $f_1, \ldots, f_m$ are functions in $\mathbb{R}^n$ into $\mathbb{R}$, then $(f_1, \ldots, f_m)$ is a function in $\mathbb{R}^n$ into $\mathbb{R}^m$. (Note. This means that (f, g) does not denote an ordered pair of functions. Fortunately we do not have to deal with ordered pairs (or ordered trios, etc.) of functions.)

Conversely, if p is a function in $\mathbb{R}^n$ into $\mathbb{R}^2$ we can define functions p_1 and p_2 by letting $p_1(x)$ and $p_2(x)$ be the first and second components respectively of $p(x)$ for each x in the domain of p. Then $p = (p_1, p_2)$.

In general, if p is in $\mathbb{R}^n$ into $\mathbb{R}^m$ we can define in this way m functions p_i such that

$$p = (p_1, \ldots, p_m).$$

These p_i are the **components** of p. Each has the same domain as p.

Exercise
5. If $f(x, y) = x^2 + y^2$, $g(x, y, z) = x + y + z$, $h(x) = (x, -x)$, $u(x) = 2x$, $v(x) = x^2 - 1$, and $w(x) = -x$ for all real numbers x, y, z, and if a subscript i denotes the ith component, specify

$$h_1, h_2, f \circ (u, v), g \circ (u, v, w), g \circ (h_1, h_1, h_2), f \circ (h_2, h_1).$$

(Note. The circle denoting composition is usually omitted if the second function is written in components. Although we wrote $f \circ (u, v)$ above, $f(u, v)$ would be more usual.)

Combinations of functions

In elementary calculus $f + g$ is defined to be the function for which $(f + g)(x) = f(x) + g(x)$, and similarly for $f - g$, fg, etc. These definitions apply automatically to functions in $\mathbb{R}^n$ into $\mathbb{R}$; we can extend them to cover functions in $\mathbb{R}^n$ into $\mathbb{R}^m$ if we define arithmetical operations in $\mathbb{R}^m$. Let us start with $\mathbb{R}^2$.

We define the sum of two number pairs by

$$(a_1, a_2) + (b_1, b_2) = (a_1 + b_1, a_2 + b_2),$$

the product of two number pairs by

$$(a_1, a_2) \cdot (b_1, b_2) = a_1 b_1 + a_2 b_2,$$

and the product of a number and a number pair by

$$c(a_1, a_2) = (ca_1, ca_2).$$

The corresponding formulae for $\mathbb{R}^3$ are

$$(a_1, a_2, a_3) + (b_1, b_2, b_3) = (a_1 + b_1, a_2 + b_2, a_3 + b_3)$$

$$(a_1, a_2, a_3) \cdot (b_1, b_2, b_3) = a_1 b_1 + a_2 b_2 + a_3 b_3$$

$$\left. \begin{array}{c} c(a_1, a_2, a_3) \\ \text{or} \quad c \cdot (a_1, a_2, a_3) \end{array} \right\} = (ca_1, ca_2, ca_3),$$

and it is clear what the definitions would be for $\mathbb{R}^m$ in general. (Note. The product of two elements of $\mathbb{R}^m$ belongs to $\mathbb{R}$, and not to $\mathbb{R}^m$. If $m = 1$, it is simply the ordinary product of two numbers.)

The identities below follow at once from the definitions. If a and b are real numbers, if α, β and γ belong to $\mathbb{R}^m$, and if Ω is the element of $\mathbb{R}^m$ whose components are all zero,

$$\alpha + \beta = \beta + \alpha \qquad\qquad (\alpha + \beta) + \gamma = \alpha + (\beta + \gamma)$$

$$\alpha \cdot \beta = \beta \cdot \alpha \qquad\qquad \alpha \cdot (\beta + \gamma) = \alpha \cdot \beta + \alpha \cdot \gamma$$

$$\alpha + \Omega = \alpha \qquad\qquad \Omega \cdot \alpha = 0$$

$$1 \cdot \alpha = \alpha \qquad\qquad a(\alpha + \beta) = a\alpha + a\beta$$

$$(a + b)\alpha = a\alpha + b\alpha \qquad\qquad a(b\alpha) = (ab)\alpha$$

$$a(\alpha \cdot \beta) = (a\alpha) \cdot \beta = \alpha \cdot (a\beta)$$

These are the 'basic laws of arithmetic' for $\mathbb{R}^m$. From them we can almost mechanically deduce similar but more complicated identities such as

$$((\alpha + \beta) + \gamma) + \delta = (\alpha + \beta) + (\gamma + \delta).$$

We use the following obvious abbreviations:

$$-\alpha \quad \text{for} \quad -1 \cdot \alpha$$
$$\alpha - \beta \quad \text{for} \quad \alpha + (-\beta)$$
$$\alpha^2 \quad \text{for} \quad \alpha \cdot \alpha$$
$$\alpha + \beta + \gamma \quad \text{for} \quad (\alpha + \beta) + \gamma.$$

Then $\alpha + \beta + \gamma$ also equals $\alpha + (\beta + \gamma)$, and this abbreviation enables us to write sums without parentheses: for example, $(\alpha + \beta) + (\gamma + \delta)$ will then be written simply $\alpha + \beta + \gamma + \delta$.

Exercise 6. What are
(a) $(1, 2) + (3, 4)$ (b) $(0, 0, 0) + (a, b, c)$
(c) $7(0, 1) - 3(2, 1)$ (d) $-(1, 0, -1) + (0, -1, 1)$
(e) $(1, 0, 2) \cdot (2, 1, 0)$ (f) $(1, 1, 1)^2$
(g) $1 + (1, 2)$ (h) $(1, 2) \cdot (1, 2, 1)$.

Problem 1. Show from the laws of arithmetic in $\mathbb{R}^m$ that $(\alpha + \beta)^2 = \alpha^2 + 2\alpha \cdot \beta + \beta^2$.

We can now define various combinations of functions. If f and g are in $\mathbb{R}^n$ into $\mathbb{R}^m$, if u is in $\mathbb{R}^n$ into $\mathbb{R}$, if $c \in \mathbb{R}$, and if $\alpha \in \mathbb{R}^m$, the functions $f + g$ etc. are defined by

$$(f + g)(\xi) = f(\xi) + g(\xi) \qquad (f + \alpha)(\xi) = f(\xi) + \alpha$$
$$(f \cdot g)(\xi) = f(\xi) \cdot g(\xi) \qquad (\alpha \cdot g)(\xi) = \alpha \cdot g(\xi)$$
$$(ug)(\xi) = u(\xi) \cdot g(\xi) \qquad (cg)(\xi) = c \cdot g(\xi)$$

for every ξ for which the right-hand side exists, in each case. We use the familiar abbreviations $f - g$ for $f + (-1)g$, f^2 for $f \cdot f$, etc. We have the obvious identities, following the pattern of the identities listed for $\mathbb{R}^m$. The following are typical:

$$f + g = g + f \qquad f + (g + h) = (f + g) + h$$
$$f \cdot (g + h) = f \cdot g + f \cdot h \qquad c(f + g) = cf + cg.$$

Their proofs are straightforward.

Exercise 7. If $\operatorname{dom} f = \operatorname{dom} g = \mathbb{R}^3$, $f(x, y, z) = (x + y, y + z)$, and $g(x, y, z) = (x - y, y - z)$, specify $f + g$, $2f - 3g$, $f \cdot g$, $f + (1, 0)$.

Problems 2. Is the inverse image under f of the image of $\mathbb{A}$ under f always equal to $\mathbb{A}$? What about the image of the inverse image?

3. Are $f \circ (g + h) = f \circ g + f \circ h$ and $(f + g) \circ h = f \circ h + g \circ h$ valid identities?

Definitions

Given a function f in $\mathbb{R}^2$ and a number a we define $f(a, \cdot)$ to be the function for which

$$f(a, \cdot)(u) = f(a, u)$$

for every u for which the right-hand side is defined. We define $f(\cdot, a)$ similarly by

$$f(\cdot, a)(u) = f(u, a).$$

We might describe $f(a, \cdot)$ as 'f with its first argument fixed at a'. We make similar definitions for functions in $\mathbb{R}^n$. □

Exercise

8. Let t be the identity function with domain $\mathbb{R}$, let (u, v) be the identity function with domain $\mathbb{R}^2$, and let (x, y, z) be the identity function with domain $\mathbb{R}^3$.
   (a) If $f = u + 2uv$, what are $f(\cdot, 0)$ and $f(1, \cdot)$?
   (b) If $f = xz + y^2$, what are $f(1, \cdot, \cdot)$ and $f(1, 2, \cdot)$?
   (c) If f is in $\mathbb{R}^3$ into $\mathbb{R}^m$, does it follow that $f(x, y, z) = f$?

2
Set theory

We shall need some concepts from set theory in our treatment of continuity.

Definition If $\mathbb{A}$ is a subset of $\mathbb{R}^m$ and $\mathbb{B}$ is a subset of $\mathbb{R}^n$, then the **cartesian product** $\mathbb{A} \times \mathbb{B}$ is

$$\{(a_1, a_2, \ldots, a_m, b_1, \ldots, b_n):(a_1, \ldots, a_m) \in \mathbb{A}$$
$$\text{and } (b_1, \ldots, b_n) \in \mathbb{B}\}.$$

Thus $\mathbb{A} \times \mathbb{B}$ is a subset of $\mathbb{R}^{m+n}$.

We define $\mathbb{A} \times \mathbb{B} \times \mathbb{C}$ to be $(\mathbb{A} \times \mathbb{B}) \times \mathbb{C}$, which clearly equals $\mathbb{A} \times (\mathbb{B} \times \mathbb{C})$, and similarly for cartesian products of four or more sets. In particular, if $\mathbb{A}_1, \ldots, \mathbb{A}_n$ are subsets of $\mathbb{R}$, then

$$\mathbb{A}_1 \times \ldots \times \mathbb{A}_n = \{(a_1, \ldots, a_n):a_i \in \mathbb{A}_i \text{ for each } i\}$$

and is a subset of $\mathbb{R}^n$. □

If $\mathbb{S}$ is a subset of $\mathbb{R}^2$ we can represent it in a diagram as a plane set. We set up axes and represent $\mathbb{S}$ by the set of points whose coordinates belong to $\mathbb{S}$. Thus if

$$\mathbb{S} = \{(x, y):x^2 + y^2 < 1\},$$

then $\mathbb{S}$ is represented by the points inside the circle with radius 1 whose centre is the origin.

Figure 2.1 shows how the cartesian product $\mathbb{A} \times \mathbb{B}$ appears if we represent $\mathbb{A}$ on one axis and $\mathbb{B}$ on the other.

Because diagrams are often used as an aid to intuition, it has become customary to blur the distinction between a number pair and the point it represents, and to refer to a number pair as a 'point'. In the same way the set $\mathbb{S}$ above may be called a 'disc' (more precisely, a disc without a rim) because that is the shape of the set that represents it. We use this terminology in higher dimensions also, and in general refer to an element of $\mathbb{R}^n$ as a point.

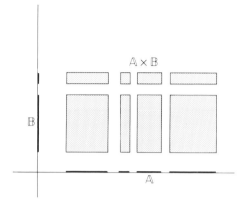

Fig. 2.1

Exercises
1. Represent $\{(x, y):0<x<1$ and $0<y<x\}$ on a diagram. Is it the cartesian product of two subsets of $\mathbb{R}$?
2. Let $\mathbb{M} = \{x:1<x^2<4\}$ and $\mathbb{N} = \{x:1<x<2\}$. Represent $\mathbb{M} \times \mathbb{N}$ in a diagram.
3. If $\mathbb{A}$, $\mathbb{B}$, and $\mathbb{C}$ each consists of two real numbers, how many elements does $\mathbb{A} \times \mathbb{B} \times \mathbb{C}$ consist of?
4. $\mathbb{S}$ is a subset of $\mathbb{R}^2$ with the following property: if (x, y) and (z, w) belong to $\mathbb{S}$, then (x, w) necessarily belong to $\mathbb{S}$. Show that $\mathbb{S}$ is the cartesian product of two subsets of $\mathbb{R}$.

Open sets

Let $\mathbb{S}$ be

$$\{(x, y):x^2 + y^2 \le 1\}.$$

Then, as we have seen, $\mathbb{S}$ is represented diagrammatically as a disc with a rim. The rim is the circle

$$\{(x, y):x^2 + y^2 = 1\},$$

and the other points are called 'interior' points. An interior point α of $\mathbb{S}$ is completely immersed in $\mathbb{S}$: there is a rectangle with centre α which is completely contained in $\mathbb{S}$ (as shown in Fig. 2.2). If α is near the rim the rectangle will have to be small, but that does not matter.

There are sets for which *every* point is interior. All we have to do to find one is to remove the rim from a set like $\mathbb{S}$. If

$$\mathbb{T} = \{(x, y):x^2 + y^2 < 1\},$$

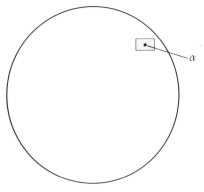

Fig. 2.2

then every point of $\mathbb{T}$ is interior to $\mathbb{T}$. Such sets are called 'open sets' and play an important role in analysis.

The difference between $\mathbb{S}$ and $\mathbb{T}$ is quite analogous to the difference (in one-dimensional calculus) between $\mathbb{A}$ and $\mathbb{B}$, where

$$\mathbb{A} = \{x : 0 \le x \le 1\}$$
$$\mathbb{B} = \{x : 0 < x < 1\}.$$

Although the difference might seem trivial from one point of view, because $\mathbb{A}$ and $\mathbb{B}$ have an infinite number of points in common and differ only in two points (0 and 1 belong to $\mathbb{A}$ and not to $\mathbb{B}$), from the point of view of the calculus this difference is important. A well-known theorem tells us that a function which is continuous on $\mathbb{A}$ is bounded on $\mathbb{A}$, whereas a function which is continuous on $\mathbb{B}$ need not be bounded on $\mathbb{B}$.

We now treat all this formally.

Definition A **neighbourhood** in $\mathbb{R}$ is a subset of the form $\{x : a < x < b\}$. This neighbourhood is denoted by $(a; b)$. (Note. The usual notation is not $(a; b)$ but (a, b). We use the semicolon to avoid confusion with the ordered pair (a, b).) □

The idea behind the name 'neighbourhood' is that a neighbourhood of x contains every number close enough to x, and a small neighbourhood of x contains *only* numbers close to x. (This will be clear when neighbourhoods are used in the theory of continuity later.)

If $\mathbb{I}$ and $\mathbb{J}$ are neighbourhoods in $\mathbb{R}$, then $\mathbb{I} \times \mathbb{J}$ is a **neighbourhood** in $\mathbb{R}^2$. More generally, if $\mathbb{I}_1, \ldots, \mathbb{I}_n$ are neighbourhoods in $\mathbb{R}$, then $I_1 \times I_2 \times \ldots \times I_n$ is a **neighbourhood** in $\mathbb{R}^n$. If ξ belongs to the neighbourhood $\mathbb{N}$, we say that $\mathbb{N}$ is a 'neighbourhood of ξ'. (Note. Diagrammatically, a neighbourhood in $\mathbb{R}^2$ is represented by the points inside a rectangle.)

Definitions

If $\mathbb{S}$ contains a neighbourhood of ξ, then ξ is **interior** to $\mathbb{S}$. The set of all points interior to $\mathbb{S}$ is the **interior** of $\mathbb{S}$. $\mathbb{S}$ is **open** if every point of $\mathbb{S}$ is interior to $\mathbb{S}$. □

Examples
1. A neighbourhood is an open set. (This is obvious.)
2. If $c \in \mathbb{R}$, then $\{x : x > c\}$ is open. (If $x > c$, choose a y between c and x, and a z greater than x. Then $(y; z)$ is a neighbourhood of x contained in the given set.)
3. $\{x : x < c\}$ is open.
4. $\{x : x \leq c\}$ and $\{x : x \geq c\}$ are not open. (The point c is not interior.)
5. The set $\mathbb{Q}$ of all points in $\mathbb{R}^2$ with rational coordinates is not open. (Any neighbourhood at all contains some points with irrational coordinates, and so no neighbourhood is contained in $\mathbb{Q}$.)
6. $\mathbb{R}^n$ is open.
7. The null set is open. (It has no non-interior points because it has no points.)

Theorem
If $\mathbb{M}$ and $\mathbb{N}$ are neighbourhoods of ξ, so is $\mathbb{M} \cap \mathbb{N}$.

Proof
In one dimension, if $\mathbb{M} = (a; b)$ and $\mathbb{N} = (c; d)$, then $a < \xi < b$ and $c < \xi < d$, and $\mathbb{M} \cap \mathbb{N} = (\max(a, c); \min(b, d))$, which is a neighbourhood of ξ.

In n dimensions, if $\mathbb{M} = \mathbb{M}_1 \times \ldots \times \mathbb{M}_n$ and $\mathbb{N} = \mathbb{N}_1 \times \ldots \times \mathbb{N}_n$, then $\mathbb{M} \cap \mathbb{N} = \mathbb{Q}_1 \times \ldots \times \mathbb{Q}_n$ where, for each i, $\mathbb{Q}_i = \mathbb{M}_i \cap \mathbb{N}_i$, which, by the result in one dimension, is a neighbourhood. Then $\mathbb{M} \cap \mathbb{N}$ is a neighbourhood; clearly it contains ξ. □

Theorem
If I_1 and I_2 are open subsets of $\mathbb{R}$, then $I_1 \times I_2$ is an open subset of $\mathbb{R}^2$. Conversely, if $I_1 \times I_2$ is open, so are I_1 and I_2.

Proof
If $(x_1, x_2) \in I_1 \times I_2$, then $x_1 \in I_1$ and $x_2 \in I_2$. Then there are neighbourhoods $\mathbb{N}_1$ and $\mathbb{N}_2$ such that

$$x_1 \in \mathbb{N}_1 \subseteq I_1 \qquad \text{and} \qquad x_2 \in \mathbb{N}_2 \subseteq I_2.$$

Then $(x_1, x_2) \in \mathbb{N}_1 \times \mathbb{N}_2 \subseteq \mathbb{I}_1 \times \mathbb{I}_2$, and $\mathbb{N}_1 \times \mathbb{N}_2$ is a neighbourhood. The converse is proved similarly. □

Corollary If $\mathbb{I}_1, \ldots, \mathbb{I}_n$ are open subsets of $\mathbb{R}$, then $\mathbb{I}_1 \times \ldots \times \mathbb{I}_n$ is an open subset of $\mathbb{R}^n$, and conversely. □

Theorem The union of any number of open sets in $\mathbb{R}^n$ is open.

Proof Any point in the union belongs to one of the given sets and so belongs to a neighbourhood contained in this set and therefore contained in the union. □

Theorem If $\mathbb{I}$ and $\mathbb{J}$ are open, so is $\mathbb{I} \cap \mathbb{J}$.

Proof If $\xi \in \mathbb{I} \cap \mathbb{J}$, then

$$\xi \in \mathbb{U} \subseteq \mathbb{I} \qquad \text{and} \qquad \xi \in \mathbb{V} \subseteq \mathbb{J},$$

where $\mathbb{U}$ and $\mathbb{V}$ are neighbourhoods. Then $\xi \in \mathbb{U} \cap \mathbb{V} \subseteq \mathbb{I} \cap \mathbb{J}$.
□

Corollary The intersection of a finite number of open sets is open. (Note. The intersection of an infinite number of open sets need not be open. For instance, the intersection of all neighbourhoods of ξ is $\{\xi\}$ and is not open.) □

Notation If $a \le b \in \mathbb{R}$, then $\{x : a \le x \le b\}$ is denoted by $[a; b]$. It differs from $(a; b)$ only inasmuch as it contains a and b. It is not open because a and b are not interior.

If $\mathbb{I}$ and $\mathbb{J}$ are sets, $\mathbb{I} \backslash \mathbb{J}$ is the set of all points of $\mathbb{I}$ that do not belong to $\mathbb{J}$ (e.g. $(0; 5) \backslash [2; 7] = (0; 2)$).

Definitions
If $\mathbb{J}$ is a subset of $\mathbb{R}^n$, its **complement** (in $\mathbb{R}^n$) is $\mathbb{R}^n \backslash \mathbb{J}$. A subset of $\mathbb{R}^n$ is **closed** if its complement is open. □

Examples 8. If $a \in \mathbb{R}$, then $\{x : x \ge a\}$ and $\{x : x \le a\}$ are closed. So is $[a; b]$ because its complement consists of two open sets. So is $\{a\}$, which is the same as $[a; a]$.
9. $\mathbb{R}^n$ is closed. (The null set is open.)
10. The null set is closed. ($\mathbb{R}^n$ is open.)
11. $[a; b] \times (c; d)$ is neither closed nor open.

Theorem If $\mathbb{I}$ and $\mathbb{J}$ are closed subsets of $\mathbb{R}$, then $\mathbb{I} \times \mathbb{J}$ is a closed subset of $\mathbb{R}^2$.

Proof If $(u, v) \notin \mathbb{I} \times \mathbb{J}$ then either $u \notin \mathbb{I}$ or $v \notin \mathbb{J}$. If $u \notin \mathbb{I}$, there is a neighbourhood $\mathbb{M}$ of u contained in the complement of $\mathbb{I}$, and so $\mathbb{M} \cap \mathbb{I} = \varnothing$. Let $\mathbb{N}$ be any neighbourhood of v. Then $(u, v) \in \mathbb{M} \times \mathbb{N}$ and

$$(\mathbb{M} \times \mathbb{N}) \cap (\mathbb{I} \times \mathbb{J}) = \varnothing;$$

therefore (u, v) is interior to the complement of $\mathbb{I} \times \mathbb{J}$. The argument is similar if $v \notin \mathbb{J}$. Therefore the complement of $\mathbb{I} \times \mathbb{J}$ is open. $\qquad \square$

Corollary If $\mathbb{I}_1, \ldots, \mathbb{I}_n$ are closed subsets of $\mathbb{R}$, then $\mathbb{I}_1 \times \ldots \times \mathbb{I}_n$ is a closed subset of $\mathbb{R}^n$. In particular, if $\alpha \in \mathbb{R}^n$, then $\{\alpha\}$ is closed.

Theorem The intersection of any number of closed sets in $\mathbb{R}^n$ is closed.

Proof Its complement is the union of the complements of the given closed sets and is therefore open. $\qquad \square$

Theorem The union of a finite number of closed sets in $\mathbb{R}^n$ is closed.

Proof Its complement is the intersection of the complements of the given closed sets and is therefore open. $\qquad \square$

Theorem If $\mathbb{I}$ is closed and $\mathbb{J}$ is open, then $\mathbb{I} \backslash \mathbb{J}$ is closed.

Proof It is the intersection of $\mathbb{I}$ with the complement of $\mathbb{J}$. $\qquad \square$

Theorem If $\mathbb{I}$ is open, its complement is closed.

Proof The complement of this complement is $\mathbb{I}$ itself and is therefore open. $\qquad \square$

Exercises 5. Show that $\{(x, y) : x^2 + y^2 < k^2\}$ is an open set, and that $\{(x, y) : x^2 + y^2 \le k^2\}$ is closed.
6. Find the intersection of all closed sets containing $(0; 1) \times (0; 1) \times (0; 1)$.
7. Prove that the intersection of all open sets containing $\mathbb{S}$ is $\mathbb{S}$ itself.

The norm

Notation If $(x, y) \in \mathbb{R}^2$, we denote $(x^2 + y^2)^{1/2}$ by

$$|(x, y)|.$$

Diagrammatically, $|(x, y)|$ is the distance from $(0, 0)$ to (x, y) and can be regarded as the 'magnitude' of (x, y). Its technical name is the **norm** (more precisely, the **Pythagorean norm**) of (x, y). If α and β belong to $\mathbb{R}^2$, $|\alpha - \beta|$ is the distance between the points α and β.

More generally, if $\alpha \in \mathbb{R}^n$, then $|\alpha|$ is

$$(\alpha_1^2 + \alpha_2^2 + \ldots + \alpha_n^2)^{1/2}.$$

It follows at once that

$$|\alpha| \geq 0$$
$$|\alpha|^2 = \alpha^2$$
$$|c\alpha| = |c| \, |\alpha| \text{ if } c \in \mathbb{R}$$
$$\text{if } |\alpha| = 0, \text{ then } \alpha = (0, 0, \ldots, 0).$$

There are two useful inequalities connected with the norm. The first is

$$|\alpha| \, |\beta| \geq \alpha \cdot \beta.$$

Proof

$$0 \leq (|\alpha| \, \beta - |\beta| \, \alpha)^2$$
$$= |\alpha|^2 \beta^2 - 2 \, |\alpha| \, |\beta| \, \alpha \cdot \beta + |\beta|^2 \alpha^2$$
$$= 2 \, |\alpha| \, |\beta| \, (|\alpha| \, |\beta| - \alpha \cdot \beta)$$

because $|\alpha|^2 = \alpha^2$ and $|\beta|^2 = \beta^2$. However, $|\alpha| \geq 0$ and $|\beta| \geq 0$, and so $|\alpha| \, |\beta| - \alpha \cdot \beta \geq 0$. $\square$

The second inequality is

$$|\alpha| + |\beta| \geq |\alpha + \beta|.$$

Proof

$$(|\alpha| + |\beta|)^2 = |\alpha|^2 + 2 \, |\alpha| \, |\beta| + |\beta|^2$$
$$\geq \alpha \cdot \alpha + 2\alpha \cdot \beta + \beta \cdot \beta$$
$$= (\alpha + \beta) \cdot (\alpha + \beta)$$
$$= |\alpha + \beta|^2.$$
$\square$

The above inequality is often called the **triangle inequality** because if we replace α and β by $\alpha - \gamma$ and $\gamma - \beta$

respectively it becomes

$$|\alpha - \gamma| + |\gamma - \beta| \geq |\alpha - \beta|. \tag{1}$$

In the usual diagrammatic representation α, β, and γ will form a triangle with sides of length $|\alpha - \gamma|$, $|\gamma - \beta|$, and $|\alpha - \beta|$. Thus (1) is equivalent to the statement that any two sides of a triangle are together greater than the third.

Definition If $\alpha \in \mathbb{R}^n$ and r is a positive number,

$$\{\xi \in \mathbb{R}^n : |\xi - \alpha| < r\}$$

is called the **disc** with centre α and radius r.

If $n = 2$, it is represented diagrammatically by the points inside a circle of radius r and centre α. Some authors use the term 'ball' instead of disc—they are thinking of the case with $n = 3$ rather than $n = 2$. □

Any neighbourhood of α contains a disc with centre α, and any disc with centre α contains a neighbourhood of α. This is obvious from a diagram if $n = 2$, because any circle with centre P contains a rectangle surrounding P and any rectangle surrounding P contains a circle with centre P. The general proof (in $\mathbb{R}^n$) is as follows.

If $\mathbb{D}$ is the disc with radius r and centre α, let $\epsilon = r/n^{1/2}$ and, for each i, let $\mathbb{N}_i$ be the neighbourhood $(\alpha_i - \epsilon; \alpha_i + \epsilon)$. Let $\mathbb{N}$ be the cartesian product of the $\mathbb{N}_i$. If ξ is any element of $\mathbb{N}$, then, successively,

$$\xi_i \in \mathbb{N}_i \qquad \text{for each } i,$$

$$|\xi_i - \alpha_i| < \epsilon,$$

$$\sum |\xi_i - \alpha_i|^2 < n\epsilon^2,$$

$$|\xi - \alpha|^2 < r^2,$$

$$\xi \in \mathbb{D}.$$

Therefore $\mathbb{N}$ is contained in $\mathbb{D}$.

Conversely, any neighbourhood $\mathbb{N}$ of α will be the cartesian product of neighbourhoods of the form $(\alpha_i - \delta_i; \alpha_i + \epsilon_i)$, where δ_i and ϵ_i are positive. Let r be the smallest of the ϵ_i and δ_i. Then the disc of radius r and centre α is contained in $\mathbb{N}$, because if

$$|\xi - \alpha| < r$$

then
$$\sum |\xi_i - \alpha_i|^2 < r^2$$

and so, for each i, $|\xi_i - \alpha_i|^2 < r^2$, whence

$$|\xi_i - \alpha_i| < r.$$

However,

$$r < \delta_i \qquad \text{and} \qquad r < \epsilon_i,$$

and so

$$\xi_i \in (\alpha_i - \delta_i, \; \alpha_i + \epsilon_i).$$

This is true for each i, and so $\xi \in \mathbb{N}$.

We can rewrite this result as follows.

Theorem If $\mathbb{N}$ is a neighbourhood of α, there is a positive number r such that
$$|\xi - \alpha| < r \qquad \text{whenever} \qquad \xi \in \mathbb{N}.$$

Conversely, if r is a positive number and $\alpha \in \mathbb{R}^n$, there is a neighbourhood $\mathbb{N}$ of α such that

$$\xi \in \mathbb{N} \qquad \text{whenever} \qquad |\xi - \alpha| < r. \qquad \square$$

We now prove that discs are open sets.

Theorem If $\mathbb{S}$ is the disc with centre α and radius r, and if β is any point of $\mathbb{S}$, then β is interior to $\mathbb{S}$.

Proof Because $\beta \in \mathbb{S}$, $|\alpha - \beta| < r$. Let $\mathbb{T}$ be the disc with centre β and radius $r - |\alpha - \beta|$. If $\xi \in \mathbb{T}$, then

$$|\xi - \beta| < r - |\alpha - \beta| \qquad (2)$$

and so

$$|\xi - \alpha| \le |\xi - \beta| + |\alpha - \beta| \quad \text{by the triangle inequality}$$
$$< r \qquad\qquad\qquad \text{by (2)}$$

and so $\xi \in \mathbb{S}$. Thus $\mathbb{T} \subseteq \mathbb{S}$. By the previous theorem, there is a neighbourhood $\mathbb{N}$ of β contained in $\mathbb{T}$. Then $\mathbb{N} \subseteq \mathbb{S}$, and so β is interior to $\mathbb{S}$. $\qquad \square$

Note We can prove similarly that $\{\xi : |\xi - \alpha| > r\}$ is an open set, and it follows that
$$\{\xi : |\xi - \alpha| \le r\}$$

is closed. This set is the disc with a rim whose centre is α and whose radius is r. Thus a disc with a rim is usually called a **closed disc** (or **closed ball**) and a disc as we have defined it is usually called an **open disc** (or **open ball**).

Some authors use the term 'neighbourhood of ξ' to mean an open disc with centre ξ. Thus when they say 'ξ has a neighbourhood contained in $\mathbb{S}$' they mean that there is an open disc with centre ξ contained in $\mathbb{S}$. The remarks on p. 14 show that if ξ has a neighbourhood contained in $\mathbb{S}$ under one definition then it has a neighbourhood contained in $\mathbb{S}$ under the other, and so the two definitions are equivalent as far as the use made of them in calculus is concerned. For example, whichever meaning is given to the word 'neighbourhood' in the definition of *interior*, the same points will turn out to be interior to a given set.

Other authors use 'neighbourhood of ξ' to mean 'open set containing ξ'. This is also equivalent to our definition.

When there is any danger of ambiguity we can refer to these last neighbourhoods as 'general neighbourhoods', to open discs as 'circular neighbourhoods', and to our neighbourhoods as 'rectangular neighbourhoods'.

Exercises

8. Find an element α of $\mathbb{R}^2$ such that $\alpha \cdot (3, 4) = 0$ and $|\alpha| = 1$. How many such elements are there?

9. Prove that if α and β are elements of $\mathbb{R}^n$, then

$$|\alpha| - |\beta| \leq |\alpha - \beta| \leq |\alpha| + |\beta|.$$

10. Let α and β be elements of $\mathbb{R}^n$. Show that the condition for the equality $|\alpha| \cdot |\beta| = \alpha \cdot \beta$ to hold is that $|\alpha| = 0$ or $|\beta| = 0$ or that there exists a positive number x for which $\beta = x\alpha$. What is the condition for each of the following equalities?

$$|\alpha| + |\beta| = |\alpha + \beta|$$

$$|\alpha| + |\beta| = |\alpha - \beta|$$

$$|\alpha| - |\beta| = |\alpha - \beta|.$$

11. Prove that $\{\xi : a < |\xi| < b\}$ is an open set.

12. Let $\mathbb{D}$ be the open disc with radius r and centre α, and let $\mathbb{D}^*$ be the open disc with radius r^* and centre α^*. Show that

$$\inf\{|\xi - \xi^*| : \xi \in \mathbb{D} \text{ and } \xi^* \in \mathbb{D}^*\} = \max\{0, |\alpha - \alpha^*| - r - r^*\}.$$

Find $\inf\{|\xi - \xi^*| : \xi \in \mathbb{N} \text{ and } \xi^* \in \mathbb{N}^*\}$, where $\mathbb{N}$ is the cartesian product of neighbourhoods $(\alpha_i - r; \alpha_i + r)$ and $\mathbb{N}^*$ is the cartesian product of neighbourhoods $(\alpha_i^* - r^*; \alpha_i^* + r^*)$.

Definition   A subset $\mathbb{S}$ of $\mathbb{R}^n$ is **bounded** if $\{|\xi| : \xi \in \mathbb{S}\}$ is bounded. $\square$

It follows immediately that open discs and neighbourhoods are bounded, subsets of bounded sets are bounded, any bounded subset of $\mathbb{R}^n$ is contained in some neighbourhood in $\mathbb{R}^n$, the union of a finite number of bounded subsets of $\mathbb{R}^n$ is bounded, and cartesian products of bounded sets are bounded.

Exercises   13. A subset of $\mathbb{R}^n$ is bounded if and only if it is a subset of some open disc with centre $(0, 0, \ldots, 0)$. Is this true or false?
14. Read through the properties of bounded sets in the last paragraph above, and sketch their formal proofs. Write out one formal proof in detail. Which of them can you prove without using the triangle inequality?

Boundaries

If $\mathbb{S}$ is a subset of $\mathbb{R}^n$, then $\mathbb{R}^n$ falls into three parts: the interior of $\mathbb{S}$, the interior of the complement of $\mathbb{S}$, which we call the **exterior** of $\mathbb{S}$, and the rest of $\mathbb{R}^n$, which we call the **boundary** of $\mathbb{S}$. Each point of $\mathbb{R}^n$ lies in just one of these parts. We denote them by $\text{int}\,\mathbb{S}$, $\text{ext}\,\mathbb{S}$ and $\text{bdy}\,\mathbb{S}$ respectively:
$\xi \in \text{int}\,\mathbb{S}$ if ξ has a neighbourhood contained in $\mathbb{S}$;
$\xi \in \text{ext}\,\mathbb{S}$ if ξ has a neighbourhood contained in the complement of $\mathbb{S}$;
$\xi \in \text{bdy}\,\mathbb{S}$ if ξ has no such neighbourhoods, in which case every neighbourhood of ξ contains both a point in $\mathbb{S}$ and a point not in $\mathbb{S}$.

Example   12. $(0; 1) \times (0; 1)$ is open, $[0; 1] \times [0; 1]$ is closed, and $[0; 1] \times (0; 1)$ is neither. All three sets have the same boundary, the same interior, and the same exterior. Diagrammatically, the boundary is a square, the interior is the set of points inside it, and the exterior is the set of points outside it.

Lemma   If $\mathbb{S} \subseteq \mathbb{T}$ and $\mathbb{T}$ is closed, then $\text{bdy}\,\mathbb{S} \subseteq \mathbb{T}$.

Proof   If $\xi \notin \mathbb{T}$, then, because $\mathbb{T}$ is closed, there is a neighbourhood $\mathbb{N}$ of ξ which does not intersect $\mathbb{T}$. Then $\mathbb{N}$ cannot intersect $\mathbb{S}$. Therefore $\xi \notin \text{bdy}\,\mathbb{S}$.

Exercises   15. $\mathbb{S}$ is the set of all rational numbers between 0 and 1. What is its boundary? What is its interior?

16. What are the boundary, interior, and exterior of

$$\{(x, y):x \geq 0 \text{ and } y \geq 0\}?$$

17. Prove that if $\mathbb{S}$ is bounded then so is its boundary.

18. Show that a set is open if and only if it contains none of its boundary points, and closed if and only if it contains all its boundary points.

19. Is the boundary of a subset of $\mathbb{S}$ necessarily a subset of the boundary of $\mathbb{S}$? Is the boundary of the union of two sets necessarily the union of the boundaries of the sets? If not, what is the relation between the boundary of the union and the union of the boundaries? Does this relation still hold for the union of an infinite number of sets? How about intersections of sets?

Compact sets

Definition   Let $\mathbb{A}$ be a set and $\mathscr{S}$ be a set of sets. Then $\mathscr{S}$ is a **covering** of $\mathbb{A}$ if the union of the sets belonging to $\mathscr{S}$ contains $\mathbb{A}$.   □

Example   13. The set of all $[x; y]$ for which $0 < x < y < 1$ is a covering of $(0; 1)$.

Definition   A covering is **open** if every set in it is open. A covering is **finite** if it consists of only a finite number of sets.   □

Example   14. The covering in the example above is neither open nor finite. However, the set of all $(x; y)$ for which $0 < x < y < 1$ is an open covering of $(0; 1)$, and $\{[0; \frac{1}{2}], [\frac{1}{2}; 1]\}$ is a finite covering of $(0; 1)$.

Note   There is no suggestion that if $\mathscr{S}$ is a covering of $\mathbb{A}$, then each set of $\mathscr{S}$ must be contained in $\mathbb{A}$. In our last example, for instance, $[0; \frac{1}{2}]$ is not contained in $(0; 1)$. As an extreme example, if $\mathbb{A}$ is any subset of $\mathbb{R}^n$, then $\{\mathbb{R}^n\}$ is a finite open covering of $\mathbb{A}$.

If $\mathscr{S}$ is a covering of $\mathbb{A}$, then in general an element of $\mathbb{A}$ might belong to more than one set of $\mathscr{S}$, and so it might be possible to remove some of the sets of $\mathscr{S}$ and still have a covering of $\mathbb{A}$. The interesting case is when $\mathscr{S}$ is infinite and we can remove all but a finite number of sets from $\mathscr{S}$ and still have a covering.

Example 15. Let $\mathscr{S}$ be the set of all open discs of radius 1 in $\mathbb{R}^2$ and $\mathbb{A}$ be $[0; 1] \times [0; 1]$. Then $\mathscr{S}$ is a covering of $\mathbb{A}$. The finite subset of $\mathscr{S}$ consisting of the open discs of radius 1 with centres at $(0, \frac{1}{2})$ and $(1, \frac{1}{2})$ is also a covering of $\mathbb{A}$. Now let $\mathscr{T}$ be the set of all $(x - 1; x + 1)$. Then $\mathscr{T}$ is a covering of $\mathbb{R}$ but no finite subset of $\mathscr{T}$ is a covering of $\mathbb{R}$. The remarkable thing is that some sets have the property that an open covering can always be reduced to a finite covering.

Definition A set $\mathbb{A}$ is **compact** if, given any open covering $\mathscr{S}$ whatsoever, a finite number of sets in $\mathscr{S}$ suffice to cover $\mathbb{A}$.
□

The Heine–Borel theorem

$[a; b]$ is compact.

Proof This is obvious if $a = b$. If $a < b$, let $\mathscr{S}$ be any open covering of $[a; b]$. Let $\mathbb{K}$ be the set of all numbers x in $[a; b]$ for which $[a; x]$ can be covered by a finite number of sets of $\mathscr{S}$. Let $\mathbb{S}$ be a set of $\mathscr{S}$ containing $\sup \mathbb{K}$. Because $\mathbb{S}$ is open, it contains a neighbourhood $(u; v)$ of $\sup \mathbb{K}$. Then there must be an element w of $\mathbb{K}$ in $\mathbb{S}$ between u and $\sup \mathbb{K}$, because if not u would be an upper bound of $\mathbb{K}$. Then $[a; w]$ can be covered by a finite number of sets of $\mathscr{S}$ and $[w; \sup \mathbb{K}]$ can be covered by one set, $\mathbb{S}$, of $\mathscr{S}$; therefore $[a; \sup \mathbb{K}]$ can be covered by a finite number of sets of $\mathscr{S}$. Thus $\sup \mathbb{K} \in \mathbb{K}$.

Clearly $\sup \mathbb{K} \leq b$. If $\sup \mathbb{K}$ were less than b, there would be a number z in $[a; b] \cap (u; v)$ greater than $\sup \mathbb{K}$. Then $[a; z]$ would be covered by a finite number of sets of $\mathscr{S}$, and consequently $z \in \mathbb{K}$. However, this is impossible because $z > \sup \mathbb{K}$.

Thus $\sup \mathbb{K} = b$ and so $b \in \mathbb{K}$, and $[a; b]$ can be covered by a finite number of sets of $\mathscr{S}$.
□

Theorem The cartesian product of n compact sets is a compact set.

Proof Let us first prove the theorem in the case $n = 2$. Then the cartesian product is of the form $\mathbb{I} \times \mathbb{J}$ where $\mathbb{I}$ and $\mathbb{J}$ are compact sets. Let $\mathscr{S}$ be any open covering of $\mathbb{I} \times \mathbb{J}$. Let x be any element of $\mathbb{I}$. Then $\mathscr{S}$ covers $\{x\} \times \mathbb{J}$. Because $\mathbb{J}$ is compact and $\{x\}$ consists of a single element, $\{x\} \times \mathbb{J}$ is compact. Therefore a finite subset $\mathscr{T}$ of $\mathscr{S}$ covers $\{x\} \times \mathbb{J}$.

Now let y be any element of $\mathbb{J}$. Then (x, y) belongs to some set $\mathbb{S}$ of $\mathcal{T}$. Because $\mathbb{S}$ is open there is a neighbourhood $\mathbb{I}_y \times \mathbb{J}_y$ such that

$$(x, y) \in \mathbb{I}_y \times \mathbb{J}_y \subseteq \mathbb{S}.$$

The set of all such $\mathbb{J}_y$ covers $\mathbb{J}$ and therefore, because $\mathbb{J}$ is compact, a finite number of them cover $\mathbb{J}$. Let $\mathbb{U}_x$ be the intersection of the corresponding $\mathbb{I}_y$. Then if (u, v) is any element of $\mathbb{U}_x \times \mathbb{J}$, v belongs to one of the finite number of $\mathbb{J}_y$ and so $(u, v) \in \mathbb{I}_y \times \mathbb{J}_y$, which is contained in the set $\mathbb{S}$ of $\mathcal{T}$. Thus $\mathcal{T}$ covers $\mathbb{U}_x \times \mathbb{J}$.

This is true for each x of $\mathbb{I}$. Because $\mathbb{I}$ is compact, a finite number of the $\mathbb{U}_x$ cover $\mathbb{I}$. However, each $\mathbb{U}_x \times \mathbb{J}$ is covered by a finite number of sets of $\mathcal{S}$. Therefore a finite number of sets of $\mathcal{S}$ cover $\mathbb{I} \times \mathbb{J}$.

This proves the case $n = 2$. In general, the given cartesian product is of the form

$$\mathbb{J}_1 \times \mathbb{J}_2 \times \ldots \times \mathbb{J}_n,$$

where each $\mathbb{J}_i$ is compact. If $\mathbb{J}_1 \times \ldots \times \mathbb{J}_r$ is compact, then so is $\mathbb{J}_1 \times \ldots \times \mathbb{J}_r \times \mathbb{J}_{r+1}$, by the result just proved. The theorem now follows by induction on r. $\qquad\square$

Definition A **compact interval** in $\mathbb{R}$ is a set of the form $[a; b]$. A **compact interval** in $\mathbb{R}^n$ is the cartesian product of n compact intervals in $\mathbb{R}$. $\qquad\square$

Theorem A subset of $\mathbb{R}^n$ is compact if and only if it is closed and bounded.

Proof First, let $\mathbb{A}$ be closed and bounded. Because $\mathbb{A}$ is bounded, it is contained in some compact interval $\mathbb{I}$. If $\mathcal{S}$ is any open covering of $\mathbb{A}$, then $\mathcal{S}$ together with the complement of $\mathbb{A}$ (which is open because $\mathbb{A}$ is closed) is an open covering of $\mathbb{I}$. A finite number of these sets cover $\mathbb{I}$ (because $\mathbb{I}$ is compact); if the complement of $\mathbb{A}$ is removed, we are left with a finite number of sets of $\mathcal{S}$ that cover $\mathbb{A}$.

Conversely, let $\mathbb{A}$ be compact. The set of all open discs of radius 1 with centres in $\mathbb{A}$ is an open covering of $\mathbb{A}$. Therefore a finite number of these discs cover $\mathbb{A}$, and so $\mathbb{A}$ is bounded.

Let $\alpha \notin \mathbb{A}$. The set of all open discs with radius r and centre ξ, for every ξ of $\mathbb{A}$ and every positive r less than

$|\xi - \alpha|$, is an open covering of $\mathbb{A}$, and so a finite set $\mathcal{T}$ of them covers $\mathbb{A}$. For the discs in $\mathcal{T}$, let

$$c = \min\{|\alpha - \xi| - r\}.$$

Because $\mathcal{T}$ is finite, this minimum exists; and it is positive. Let η be any element of $\mathbb{A}$. It will lie in some open disc of $\mathcal{T}$ and, if ξ is the centre and r the radius of this disc,

$$|\xi - \eta| < r. \tag{3}$$

Then

$$|\alpha - \eta| \geq |\alpha - \xi| - |\xi - \eta|$$
$$> |\alpha - \xi| - r \quad \text{by (3)}$$
$$\geq c.$$

Therefore η does not lie in the open disc with centre α and radius c. Consequently, this open disc contains no elements of $\mathbb{A}$: it is contained in the complement of $\mathbb{A}$. It follows that α is an interior point of this complement. However, α was any point of the complement; therefore the complement is open and $\mathbb{A}$ itself is closed. $\qquad\square$

Exercises
20. Is a finite set necessarily compact?
21. Is the union of two compact sets necessarily compact? How about the union of an infinite number of compact sets?
22. Is a subset of a compact set necessarily compact?
23. Is the intersection of two compact sets necessarily compact? How about the intersection of an infinite number of compact sets?

Problem
1. If $\mathbb{A}$ is compact, $\mathbb{B}$ is open, and $\mathbb{A} \subseteq \mathbb{B}$, does there necessarily exist a compact set $\mathbb{C}$ such that $\mathbb{A} \subseteq \text{int } \mathbb{C}$ and $\mathbb{C} \subseteq \mathbb{B}$?

3
Continuity

The idea of continuity in many dimensions is the same as that of continuity in one dimension.

Definition A function f in $\mathbb{R}^n$ into $\mathbb{R}^m$ is **continuous** at a point α of $\text{dom} f$ if, for each neighbourhood $\mathbb{N}$ of $f(\alpha)$, there is a neighbourhood $\mathbb{U}$ of α such that $f(\xi) \in \mathbb{N}$ whenever $\xi \in \mathbb{U}$.

□

It follows at once that f cannot be continuous at α unless α is interior to $\text{dom} f$. An identity function or constant function is obviously continuous at any point interior to its domain.

Theorem Let f be a function in $\mathbb{R}^n$ into $\mathbb{R}^m$, and let $\alpha \in \text{dom} f$. Then f is continuous at α if and only if each component is continuous at α.

Proof We first suppose that f is continuous at α. Let $\mathbb{N}_i$ be any neighbourhood of $f_i(\alpha)$. Let $\mathbb{N}$ be any neighbourhood of $f(\alpha)$ whose ith component is $\mathbb{N}_i$. There is a neighbourhood $\mathbb{U}$ of α such that

$$f(\xi) \in \mathbb{N} \qquad \text{whenever} \qquad \xi \in \mathbb{U}$$

and then $f_i(\xi) \in \mathbb{N}_i$. Therefore f_i is continuous at α. This is true for each i. Conversely, suppose that each f_i is continuous at α. Let $\mathbb{N}$ be any neighbourhood of $f(\alpha)$. It will be of the form

$$\mathbb{N}_1 \times \mathbb{N}_2 \times \ldots \times \mathbb{N}_m.$$

For each i there is a neighbourhood $\mathbb{U}_i$ of $f_i(\alpha)$ such that

$$f_i(\xi) \in \mathbb{N}_i \qquad \text{whenever} \qquad \xi \in \mathbb{U}_i.$$

Let $\mathbb{U}$ be the intersection of all the $\mathbb{U}_i$. Then

$$f(\xi) \in \mathbb{N} \qquad \text{whenever} \qquad \xi \in \mathbb{U}.$$

Therefore f is continuous at α.

□

The following criterion for continuity is so close to the definition that it can be regarded as an alternative definition. We shall use it in some of the upcoming theorems.

The ϵ, δ criterion for continuity

Let f be in $\mathbb{R}^n$ into $\mathbb{R}^m$ and $\alpha \in \mathbb{R}^n$. Then f is continuous at α if and only if, for each positive ϵ, there is a positive δ such that

$$|f(\xi) - f(\alpha)| < \epsilon \quad \text{whenever} \quad |\xi - \alpha| < \delta.$$

Proof Suppose first that f is continuous at α. Let ϵ be any positive number. There is a neighbourhood $\mathbb{N}$ of $f(\alpha)$ such that

$$|\tau - f(\alpha)| < \epsilon \quad \text{whenever} \quad \tau \in \mathbb{N}. \tag{1}$$

By definition of continuity, there is a neighbourhood $\mathbb{U}$ of α such that

$$f(\xi) \in \mathbb{N} \quad \text{whenever} \quad \xi \in \mathbb{U}. \tag{2}$$

There is a positive δ such that

$$\xi \in \mathbb{U} \quad \text{whenever} \quad |\xi - \alpha| < \delta. \tag{3}$$

By (2) and (3), $f(\xi) \in \mathbb{N}$ whenever $|\xi - \alpha| < \delta$ and then, by (1),

$$|f(\xi) - f(\alpha)| < \epsilon.$$

The converse is proved equally easily. $\square$

Example 1. Let $f(x, y) = 2x + 3y$ for every (x, y) in $\mathbb{R}^2$, and let $\alpha = (a, b)$. We can use the ϵ, δ criterion to show that f is continuous at α as follows. Let ϵ be any positive number. Then there is a positive number δ, namely $\epsilon/5$, such that

$$|f(x, y) - f(a, b)| < \epsilon \quad \text{whenever} \quad |(x, y) - (a, b)| < \delta.$$

Proof If $|(x, y) - (a, b)| < \epsilon/5$, then both $|x - a| < \epsilon/5$ and $|y - b| < \epsilon/5$. Then

$$|f(x, y) - f(a, b)| = |2x + 3y - 2a - 3b|$$
$$\leq 2|x - a| + 3|y - b|$$
$$< \frac{2\epsilon}{5} + \frac{3\epsilon}{5}$$
$$= \epsilon. \qquad \square$$

There is a group of basic theorems which say, in effect, that a mathematical combination of continuous functions is continuous. More precisely, let f and g be functions in $\mathbb{R}^n$ into $\mathbb{R}^m$ that are continuous at α. Let $k \in \mathbb{R}$ and $\beta \in \mathbb{R}^m$. Then the functions $f + g$, $f + \beta$, kf, $\beta \cdot f$ (the product in $\mathbb{R}^m$) and $f \cdot g$ are continuous at α. If $m = 1$, so that $1/f$ is defined, then $1/f$ is continuous at α provided that $f(\alpha) \neq 0$. Finally, the composite of continuous functions is continuous. Let us now prove these theorems.

Theorem If f and g are functions in $\mathbb{R}^n$ into $\mathbb{R}^m$ that are continuous at α, then $f + g$ is continuous at α.

Proof Let ϵ be any positive number. By the criterion for continuity there is a δ_1 such that

$$|f(\xi) - f(\alpha)| < \tfrac{1}{2}\epsilon \qquad \text{whenever} \qquad |\xi - \alpha| < \delta_1$$

and there is a δ_2 such that

$$|g(\xi) - g(\alpha)| < \tfrac{1}{2}\epsilon \qquad \text{whenever} \qquad |\xi - \alpha| < \delta_2.$$

Let $\delta = \min(\delta_1, \delta_2)$. Then, whenever $|\xi - \alpha| < \delta$, we have both

$$|f(\xi) - f(\alpha)| < \tfrac{1}{2}\epsilon$$

and

$$|g(\xi) - g(\alpha)| < \tfrac{1}{2}\epsilon$$

and so

$$|f(\xi) + g(\xi) - f(\alpha) - g(\alpha)| < \epsilon,$$

that is

$$|(f + g)(\xi) - (f + g)(\alpha)| < \epsilon.$$

Consequently $f + g$ is continuous at α. $\qquad\square$

Theorem If f is a function in $\mathbb{R}^n$ into $\mathbb{R}^m$ that is continuous at α and if $\beta \in \mathbb{R}^m$, then the function $f + \beta$ is continuous at α.

Proof Let $g(\xi) = \beta$ for every ξ in $\mathbb{R}^n$. Then g is continuous at α and $f + g = f + \beta$. The result follows by the previous theorem. $\qquad\square$

Corollaries (continuity of a sum)

Let f, g, h, k, etc. be functions in $\mathbb{R}^n$ into $\mathbb{R}^m$, and let $\beta \in \mathbb{R}^m$. If the named functions are continuous at α, then so are

$$f + g + h, \qquad f + g + \beta, \qquad f + g + h + k, \qquad \text{etc.}$$

In fact the sum of any (finite) number of continuous functions and constants is continuous. $\qquad\square$

Theorem   If f is a function in $\mathbb{R}^n$ into $\mathbb{R}^m$ which is continuous at α and if $k \in \mathbb{R}$, then $k \cdot f$ is continuous at α.

Proof   If $k = 0$ the result follows because $k \cdot f$ is constant. If $k \neq 0$, let ϵ be any positive number. Then there is a δ such that

$$|f(\xi) - f(\alpha)| < \epsilon / |k|$$

whenever $|\xi - \alpha| < \delta$. Then

$$|k \cdot f(\xi) - k \cdot f(\alpha)| < \epsilon$$

whenever $|\xi - \alpha| < \delta$. Consequently, $k \cdot f$ is continuous at α. $\square$

Theorem   If f is a function in $\mathbb{R}^n$ into $\mathbb{R}^m$ which is continuous at α, and if $\beta \in \mathbb{R}^m$, then the function $\beta \cdot f$ is continuous at α.

Proof   Each component f_i of f is continuous at α. Therefore so is each $\beta_i \cdot f_i$. Therefore so is their sum, which is $\beta \cdot f$. $\square$

Theorem (continuity of a product)

If f and g are functions in $\mathbb{R}^n$ into $\mathbb{R}^m$ that are continuous at α, then the function $f \cdot g$ is continuous at α.

Proof   Set $u = f - f(\alpha)$ and $v = g - g(\alpha)$. Then $u(\alpha) = v(\alpha) = 0$ and (by the continuity of a sum) u and v are continuous at α. By the previous theorem, so are $f(\alpha) \cdot v$ and $g(\alpha) \cdot u$. Let ϵ be any positive number. There are positive numbers δ_1 and δ_2 such that

$$|u(\xi)| < \epsilon^{1/2} \quad \text{whenever} \quad |\xi - \alpha| < \delta_1$$

and

$$|v(\xi)| < \epsilon^{1/2} \quad \text{whenever} \quad |\xi - \alpha| < \delta_2.$$

Let $\delta = \min(\delta_1, \delta_2)$. Then

$$|u(\xi) \cdot v(\xi)| < \epsilon \quad \text{whenever} \quad |\xi - \alpha| < \delta.$$

Consequently $u \cdot v$ is continuous at α. But

$$f \cdot g = (u + f(\alpha)) \cdot (v + g(\alpha))$$
$$= u \cdot v + f(\alpha) \cdot v + g(\alpha) \cdot u + f(\alpha) \cdot g(\alpha)$$

and so, by the continuity of a sum, $f \cdot g$ is continuous at α. $\square$

Corollary   If F is a function in $\mathbb{R}^n$ into $\mathbb{R}$ and g is a function in $\mathbb{R}^n$ into $\mathbb{R}^m$, then, if F and g are continuous at α, so is the function $F \cdot g$.

Proof   The ith component of $F \cdot g$ is $F \cdot g_i$, which is continuous at α by the previous theorem with $m = 1$.

Theorem   If f is a function in $\mathbb{R}^n$ into $\mathbb{R}$ which is continuous at α, and if $f(\alpha) \neq 0$, then $1/f$ is continuous at α.

Proof   Let $\mathbb{N}$ be any neighbourhood of $1/f(\alpha)$. It will contain a neighbourhood $(b; c)$ of $1/f(\alpha)$ for which b and c are of the same sign. Then $(1/c; 1/b)$ is a neighbourhood of $f(\alpha)$. Therefore there is a neighbourhood $\mathbb{U}$ of α such that

$$f(\xi) \in (1/c; 1/b)$$

whenever $\xi \in \mathbb{U}$. But then

$$(1/f)(\xi) \in (b; c) \subseteq \mathbb{N}.$$

Consequently $1/f$ is continuous at α. $\square$

Theorem (continuity of a composite)

If f is continuous at α and h is continuous at $f(\alpha)$, then $h \circ f$ is continuous at α.

Proof   Let $\mathbb{U}$ be any neighbourhood of $h(f(\alpha))$. Then there is a neighbourhood $\mathbb{V}$ of $f(\alpha)$ such that

$$h(\xi) \in \mathbb{U} \qquad \text{whenever} \qquad \xi \in \mathbb{V},$$

and there is a neighbourhood $\mathbb{W}$ of α such that

$$f(\eta) \in \mathbb{V} \qquad \text{whenever} \qquad \eta \in \mathbb{W}.$$

Then

$$h(f(\eta)) \in \mathbb{U} \qquad \text{whenever} \qquad \eta \in \mathbb{W}.$$ $\square$

Example   2. Let $f(x, y) = xy/(x^2 + y^2)$ whenever $(x, y) \neq (0, 0)$, and let $f(0, 0) = 0$. Prove that f is not continuous at $(0, 0)$.

Solution   If f were continuous at $(0, 0)$ then, by the ϵ, δ criterion, there would be a positive number δ such that

$$|f(x, y) - f(0, 0)| < 1/100$$

$$\text{whenever} \qquad |(x, y) - (0, 0)| < \delta. \quad (4)$$

(We have taken the ϵ in the criterion to be $1/100$, which we are entitled to do because the criterion holds for every positive ϵ.) Take $x = y = \frac{1}{2}\delta$. Then on the one hand

$$|(x, y) - (0, 0)| = (\tfrac{1}{4}\delta^2 + \tfrac{1}{4}\delta^2)^{1/2} = \frac{\delta}{2^{1/2}}$$

which is less than δ, but on the other hand

$$|f(x, y) - f(0, 0)| = \frac{\tfrac{1}{4}\delta^2}{(\tfrac{1}{4}\delta^2 + \tfrac{1}{4}\delta^2)}$$

$$= \tfrac{1}{2}$$

which is *not* less than $1/100$. Therefore (4) does not hold.

Example 2 (continued)
If $(a, b) \neq (0, 0)$, prove that f is continuous at (a, b).

Solution Let $g(x, y) = x$, $h(x, y) = x^2$, $k(x, y) = y$, $m(x, y) = y^2$, $p(x, y) = x^2 + y^2$ and $q(x, y) = xy$ for every (x, y) in $\mathbb{R}^2$. Then g and k are continuous at (a, b) because they are components of the identity function on $\mathbb{R}^2$, which is continuous there. Then $h = g^2$ and $m = k^2$, and so h and m are continuous at (a, b) by the continuity of a product. Then, because $p = h + m$, p is continuous at (a, b) by the continuity of a sum. Because $q = gk$, q is continuous at (a, b) by the continuity of a product. Finally $f = q/p$ on a neighbourhood of (a, b) and $p(a, b) \neq 0$, and so f is continuous at (a, b).

Note We can sum this up by saying that, except at $(0, 0)$, f is an algebraic combination of continuous functions and so, by the appropriate basic theorems, f is continuous. We would in fact normally say this, and not go through all the details above.

Because (as we know from elementary analysis) the square-root function is continuous, we can use square roots in the combinations. For example, the function f on $\mathbb{R}^{2n}$ into $\mathbb{R}$ defined by

$$f(\xi_1, \ldots, \xi_n, \eta_1, \ldots, \eta_n) = \left\{ \sum_{i=1}^{n} (\xi_i - \eta_i)^2 \right\}^{1/2}$$

for every ξ and η in $\mathbb{R}^n$ is continuous. (We shall use this result later.)

Exercises   1. Let $f(x, y) = x^3/(x^2 + y^2)$ whenever $(x, y) \neq (0, 0)$,   and $f(0, 0) = 0$. Is f continuous at $(0, 0)$? Prove from the definition that your answer is correct.

2. As Exercise 1, but with $x^2 y/(x^2 + y^2)$ in place of $x^3/(x^2 + y^2)$.

3. Let $f(x, y) = x^3/(x^2 - y^2)$ whenever $x^2 \neq y^2$ and $f(x, y) = c$ whenever $x^2 = y^2$. Is there a value of c for which f is continuous at $(0, 0)$?

4. Let $f(x, y) = x^2/(x^2 + y^2)$ whenever $(x, y) \neq (0, 0)$ and $f(0, 0) = c$. Is there a value of c for which f is continuous at $(0, 0)$?

5. Let $p \geq 0$, $q \geq 0$, and $p + q > 2$. Prove that, if $f(0, 0) = 0$ and $f(x, y) = x^p y^q/(x^2 + y^2)$ whenever $(x, y) \neq (0, 0)$, then f is continuous at $(0, 0)$.

6. Let $f(x, y, z) = xyz/(x^2 + y^2 + z^2)$ whenever $(x, y, z) \neq (0, 0, 0)$, and $f(0, 0, 0) = c$. Is there a value of c for which f is continuous at $(0, 0, 0)$?

7. We stated that obviously an identity function is continuous at every point interior to its domain. It will do no harm to prove this formally from the definition. Why does it follow immediately that if $f(x, y) = x$ for every (x, y) in $\mathbb{R}^2$, then f is continuous everywhere?

8. Let $f(x, y) = x^2 y/(x^2 + y^2)$ whenever $(x, y) \neq (0, 0)$,   and $f(0, 0) = 0$. Then Exercise 2 shows that f is continuous at $(0, 0)$. Show that f is continuous everywhere else. Then show that if $g(x, y) = x^3 y^2/(x^2 + y^2)$ whenever $(x, y) \neq (0, 0)$ and $g(0, 0) = 0$, it follows that g is continuous everywhere.

9. Let $f(x, y) = x^3/(x^2 - y^2)$ whenever $x^2 \neq y^2$,   and $f(x, y) = 0$ whenever $x^2 = y^2$. Where is f continuous?

10. Let $c > 0$. Prove that if $f(\xi) = g(\xi)$ whenever $0 < |\xi - \alpha| < c$, and f and g are continuous at α, then $f(\alpha) = g(\alpha)$.

Problems   1. Is it true that f is continuous at α if and only if, for each i, $f(\alpha_1, \alpha_2, \ldots, \alpha_{i-1}, \cdot, \alpha_{i+1}, \ldots, \alpha_n)$ is continuous at α_i?

2. What can be said about the continuity of the inverse of a continuous function?

3. $f(x, y) = (g_1(x, 0), g_2(0, y))$ for every x and y in $\mathbb{R}$. If f is continuous at (a, b), does it follow that g is continuous at (a, b)? If g is continuous at (a, b) does it follow that f is continuous at (a, b)? If f is continuous everywhere, does it follow that g is continuous everywhere? If g is continuous everywhere, does it follow that f is continuous everywhere?

Continuity on sets

A function f cannot be continuous at α unless α is interior to dom f. Consequently, f cannot be continuous everywhere in dom f unless dom f is open, no matter how well behaved f may be. We therefore make the following definitions.

Definitions

Let $\mathbb{A} \subseteq \text{dom } f$ and $\alpha \in \mathbb{A}$. Then f is **continuous on** $\mathbb{A}$ **at** α if, for each neighbourhood $\mathbb{N}$ of $f(\alpha)$, there is a neighbourhood $\mathbb{U}$ of α such that

$$f(\xi) \in \mathbb{N} \qquad \text{whenever} \qquad \xi \in \mathbb{U} \cap \mathbb{A}.$$

(Notice the '$\xi \in \mathbb{U} \cap \mathbb{A}$' which occurs where '$\xi \in \mathbb{U}$' occurs in the straightforward definition of 'continuous at α'.)

f is **continuous on** $\mathbb{A}$ if it is continuous on $\mathbb{A}$ at α for each α of $\mathbb{A}$.

f is **continuous** if it is continuous on its domain. $\qquad \square$

Examples
3. Constant functions and identity functions are continuous.
4. If α is interior to $\mathbb{A}$, f is continuous on $\mathbb{A}$ at α if and only if it is continuous at α.
5. If $\mathbb{A}$ is open, f is continuous on $\mathbb{A}$ if and only if it is continuous at each point of $\mathbb{A}$.
6. If g is continuous on $\mathbb{A}$ and f is the restriction of g to a set containing $\mathbb{A}$, then f is continuous on $\mathbb{A}$.
7. If there is a neighbourhood $\mathbb{Q}$ of α such that α is the *only* point of $\mathbb{A}$ in $\mathbb{Q}$, then f is (trivially) continuous on $\mathbb{A}$ at α.
8. If $f(x, y) = (1 - x^2 - y^2)^{1/2}$ for every (x, y) of $\mathbb{R}^2$ for which $(1 - x^2 - y^2)^{1/2}$ exists, and if this set of (x, y) is the domain of f, then f is continuous.

Let f and g both be continuous on $\mathbb{A}$ at α. Then we would expect $f + g$ to be continuous on $\mathbb{A}$ at α. This is in fact true, and its proof is like the proof of the theorem that if f and g are continuous at α then so is $f + g$. The only difference is that we must insert the phrase 'and $\xi \in \mathbb{A}$' at appropriate points. The same applies to other combinations of functions. Let us state all this formally.

Theorem

Let f and g be functions in $\mathbb{R}^n$ into $\mathbb{R}^m$, and let F be in $\mathbb{R}^n$ into $\mathbb{R}$. Let $c \in \mathbb{R}$ and $\beta \in \mathbb{R}^m$. Then if f, g and F are continuous on $\mathbb{A}$ at α, so are $f + g$, $f + \beta$, $f \cdot g$, $\beta \cdot f$, $c \cdot f$ and $F \cdot f$. So is $1/F$ if $F(\alpha) \neq 0$.

Proof

Let ϵ be any positive number. Then there are positive numbers δ_1 and δ_2 such that

$$|f(\xi) - f(\alpha)| < \tfrac{1}{2}\epsilon$$

whenever
$$|\xi - \alpha| < \delta_1 \quad \text{and} \quad \xi \in \mathbb{A},$$
and
$$|g(\xi) - g(\alpha)| < \tfrac{1}{2}\epsilon$$
whenever
$$|\xi - \alpha| < \delta_2 \quad \text{and} \quad \xi \in \mathbb{A}.$$
Let δ be the smaller of δ_1 and δ_2. Then
$$|f(\xi) + g(\xi) - f(\alpha) - g(\alpha)| < \epsilon$$
whenever
$$|\xi - \alpha| < \delta \quad \text{and} \quad \xi \in \mathbb{A}.$$
Therefore $f + g$ is continuous on $\mathbb{A}$ at α.

The proofs of the other parts of the theorem are obtained in the same way. $\square$

We have two useful criteria for continuity on $\mathbb{A}$ in terms of open sets and inverse images.†

Theorem Let f be in $\mathbb{R}^n$ into $\mathbb{R}^m$ and $\mathbb{A} \subseteq \operatorname{dom} f$. Then f is continuous on $\mathbb{A}$ if and only if, for each open set $\mathbb{W}$ in $\mathbb{R}^m$, each α of $(f^{\frown}\mathbb{W}) \cap \mathbb{A}$ has a neighbourhood $\mathbb{U}_\alpha$ such that
$$\mathbb{U}_\alpha \cap \mathbb{A} \subseteq f^{\frown}\mathbb{W}.$$

Proof Let f be continuous on $\mathbb{A}$ and $\mathbb{W}$ be open. Then f is continuous on $\mathbb{A}$ at each α of $(f^{\frown}\mathbb{W}) \cap \mathbb{A}$. Because $f(\alpha) \in \mathbb{W}$, $\mathbb{W}$ contains a neighbourhood $\mathbb{N}_\alpha$ of $f(\alpha)$. Then there is a neighbourhood $\mathbb{U}_\alpha$ of α such that
$$f(\xi) \in \mathbb{N}_\alpha \quad \text{whenever} \quad \xi \in \mathbb{U}_\alpha \cap \mathbb{A}.$$
Then $f(\xi) \in \mathbb{W}$ and so $\xi \in f^{\frown}\mathbb{W}$. Consequently
$$\mathbb{U}_\alpha \cap \mathbb{A} \subseteq f^{\frown}\mathbb{W}.$$

Conversely, suppose that there are neighbourhoods $\mathbb{U}_\alpha$ as described. Let α be any point of $\mathbb{A}$ and $\mathbb{N}$ any neighbourhood of $f(\alpha)$. Then there is a neighbourhood $\mathbb{U}_\alpha$ of α such that
$$\mathbb{U}_\alpha \cap \mathbb{A} \subseteq f^{\frown}\mathbb{N}.$$
Thus $f(\xi) \in \mathbb{N}$ whenever $\xi \in \mathbb{U}_\alpha \cap \mathbb{A}$. Therefore f is continuous on $\mathbb{A}$ at α. $\square$

† Reminder: we denote the inverse image of $\mathbb{W}$ under f by $f^{\frown}\mathbb{W}$ (see p. 2).

Theorem Let f be in $\mathbb{R}^n$ into $\mathbb{R}^m$ and $\mathbb{A} \subseteq \mathrm{dom} f$. Then f is continuous on $\mathbb{A}$ if and only if, for each open set $\mathbb{W}$ in $\mathbb{R}^m$, there is an open set $\mathbb{V}$ such that

$$(f^{\sim}\mathbb{W}) \cap \mathbb{A} = \mathbb{V} \cap \mathbb{A}.$$

Proof Let f be continuous on $\mathbb{A}$ and $\mathbb{W}$ be open. If $(f^{\sim}\mathbb{W}) \cap \mathbb{A} = \varnothing$, then $\varnothing$ will do for $\mathbb{V}$. If not, then by the previous theorem, each α of $(f^{\sim}\mathbb{W}) \cap \mathbb{A}$ has a neighbourhood $\mathbb{U}_\alpha$ such that

$$\mathbb{U}_\alpha \cap \mathbb{A} \subseteq f^{\sim}\mathbb{W}.$$

Let $\mathbb{V}$ be the union of the $\mathbb{U}_\alpha$ for every α in $(f^{\sim}\mathbb{W}) \cap \mathbb{A}$. Then $\mathbb{V}$ is open and

$$\mathbb{V} \cap \mathbb{A} \subseteq (f^{\sim}\mathbb{W}) \cap \mathbb{A}.$$

We now prove the reverse inequality. If

$$\alpha \in (f^{\sim}\mathbb{W}) \cap \mathbb{A}$$

then

$$\alpha \in \mathbb{U}_\alpha \subseteq \mathbb{V} \quad \text{and} \quad \alpha \in \mathbb{A},$$

and therefore

$$(f^{\sim}\mathbb{W}) \cap \mathbb{A} \subseteq \mathbb{V} \cap \mathbb{A}.$$

Conversely, suppose that the $\mathbb{V}$s exist as described. Let α be any point of $\mathbb{A}$ and $\mathbb{N}$ be any neighbourhood of $f(\alpha)$. Then there is a $\mathbb{V}$ such that

$$(f^{\sim}\mathbb{N}) \cap \mathbb{A} = \mathbb{V} \cap \mathbb{A}.$$

Then $\alpha \in \mathbb{V}$ and so $\mathbb{V}$ contains a neighbourhood $\mathbb{U}_\alpha$ of α:

$$\mathbb{U}_\alpha \cap \mathbb{A} \subseteq \mathbb{V} \cap \mathbb{A} \subseteq f^{\sim}\mathbb{N}.$$

Therefore f is continuous on $\mathbb{A}$ at α. □

Corollary (i)

f is continuous if and only if for each open set $\mathbb{W}$ in $\mathbb{R}^m$ there is an open set $\mathbb{V}$ such that

$$f^{\sim}\mathbb{W} = \mathbb{V} \cap \mathrm{dom} f.$$

Proof Set $\mathbb{A} = \mathrm{dom} f$ and note that $f^{\sim}\mathbb{W} \subseteq \mathrm{dom} f$. □

Corollary (ii)

A function with open domain is continuous if and only if the inverse image of every open set is open.

Proof   If f is continuous, $f^{\sim}\mathbb{W} = \mathbb{V} \cap \operatorname{dom} f$ where $\mathbb{V}$ and $\operatorname{dom} f$ are both open. Conversely, if $f^{\sim}\mathbb{W}$ is open, set $\mathbb{V} = f^{\sim}\mathbb{W}$.   □

The theorem and its corollaries have counterparts for closed sets. Let us prove as a theorem the counterpart of corollary (i).

Theorem   Let f be a function from $\mathbb{R}^n$ to $\mathbb{R}^m$. Then f is continuous if and only if, whenever $\mathbb{U}$ is a closed subset of $\mathbb{R}^m$, there is a closed subset $\mathbb{Q}$ of $\mathbb{R}^n$ such that

$$f^{\sim}\mathbb{U} = \mathbb{Q} \cap \operatorname{dom} f.$$

Proof   If $\mathbb{U}$ is a closed subset of $\mathbb{R}^m$, then $\mathbb{R}^m \setminus \mathbb{U}$ is open. Therefore, if f is continuous, there is an open set $\mathbb{N}$ such that

$$f^{\sim}(\mathbb{R}^m \setminus \mathbb{U}) = \mathbb{N} \cap \operatorname{dom} f.$$

Then

$$\begin{aligned}
f^{\sim}\mathbb{U} &= \operatorname{dom} f \setminus f^{\sim}(\mathbb{R}^m \setminus \mathbb{U}) \\
&= \operatorname{dom} f \setminus (\mathbb{N} \cap \operatorname{dom} f) \\
&= (\mathbb{R}^n \setminus \mathbb{N}) \cap \operatorname{dom} f
\end{aligned}$$

and $\mathbb{R}^n \setminus \mathbb{N}$ is closed.   □

Corollary   Let f be a function from $\mathbb{R}^n$ to $\mathbb{R}^m$. If $f^{\sim}\mathbb{U}$ is closed whenever $\mathbb{U}$ is a closed subset of $\mathbb{R}^m$, then f is continuous.

Proof $f^{\sim}\mathbb{U} \subseteq \operatorname{dom} f$;   therefore $f^{\sim}\mathbb{U} = f^{\sim}\mathbb{U} \cap \operatorname{dom} f = \mathbb{Q} \cap \operatorname{dom} f$ where $\mathbb{Q}$ $(= f^{\sim}\mathbb{U})$ is closed.   □

Exercises   11. Prove that if f is continuous on $\mathbb{A}$ and $\mathbb{B} \subseteq \mathbb{A}$, then f is continuous on $\mathbb{B}$.

12. Let f be continuous at α. Let $\mathbb{N}$ be a neighbourhood of $f(\alpha)$. Show that $f^{\sim}\mathbb{N}$ contains a neighbourhood of α.

13. Let $[a; b] \subseteq \operatorname{dom} f$, where $a < b$. Prove that f is continuous on $[a; b]$ if and only if it is continuous from above at a and from below at b and continuous at all numbers between a and b. (f is continuous from above at a if, given any neighbourhood $\mathbb{N}$ of $f(a)$, there is a neighbourhood $\mathbb{U}$ of a such that $f(x) \in \mathbb{N}$ whenever x lies in $\mathbb{U}$ and is greater than a. A similar argument holds for continuity from below.)

14. If f is continuous on $f^{\sim}\mathbb{U}$ and $\mathbb{U}$ is open, must $f^{\sim}\mathbb{U}$ be open?

15. If f is continuous on $f^{\sim}\mathbb{U}$ and $f^{\sim}\mathbb{U}$ is open, must $\mathbb{U}$ be open?

One of the most important properties of compact sets is that their images under continuous functions are compact.

Theorem Let $\mathbb{A}$ be compact and f be continuous on $\mathbb{A}$. Then $f(\mathbb{A})$ is compact.

Proof Let $\mathscr{S}$ be any open covering of $f(\mathbb{A})$. For each set $\mathbb{S}$ of this covering there is an open set $\mathbb{V}$ for which $\mathbb{V} \cap \mathbb{A} = f^{-}\mathbb{S} \cap \mathbb{A}$. The set of all the $f^{-}\mathbb{S}$ cover $\mathbb{A}$; therefore the $\mathbb{V}$s do so, and therefore a finite number of $\mathbb{V}$s do so. The corresponding $\mathbb{S}$s cover $f(\mathbb{A})$. $\qquad\square$

Corollary If f is in $\mathbb{R}^n$ into $\mathbb{R}$ and is continuous on a compact set $\mathbb{A}$, then there are elements α and β of $\mathbb{A}$ such that

$$f(\alpha) = \sup f(\mathbb{A}) \qquad \text{and} \qquad f(\beta) = \inf f(\mathbb{A}).$$

Proof $f(\mathbb{A})$ is compact and therefore bounded. Therefore $\sup f(\mathbb{A})$ exists: call it σ. If there is no α in $\mathbb{A}$ for which

$$f(\alpha) = \sigma$$

let

$$g = \frac{1}{\sigma - f}.$$

Then g is continuous on $\mathbb{A}$ and so g is bounded on $\mathbb{A}$. Let τ be an upper bound. Then $\tau > 0$ and

$$0 < \{\sigma - f(\xi)\}^{-1} \le \tau \qquad \text{whenever} \qquad \xi \in \mathbb{A}$$

and so

$$f(\xi) \le \sigma - \tau^{-1} \qquad \text{whenever} \qquad \xi \in \mathbb{A}$$

which contradicts the fact that σ is the *least* upper bound of f on $\mathbb{A}$. Thus there must be an α as required; the same applies to β. $\qquad\square$

Theorem If $\mathbb{A}$ and $\mathbb{B}$ are non-null compact subsets of $\mathbb{R}^n$, there is a point α of $\mathbb{A}$ and a point β of $\mathbb{B}$ such that

$$|\xi - \eta| \ge |\alpha - \beta| \qquad \text{whenever} \qquad \xi \in \mathbb{A} \text{ and } \eta \in \beta. \quad (5)$$

Proof $\mathbb{A} \times \mathbb{B}$ is compact by the theorem on p. 19. The function f on $\mathbb{R}^{2n}$ into $\mathbb{R}$ for which

$$f(\xi_1, \ldots, \xi_n, \eta_1, \ldots, \eta_n) = |\xi - \eta|$$

for every ξ and η of $\mathbb{R}^n$

is continuous (see p. 27) and so has a minimum on $\mathbb{A} \times \mathbb{B}$. $\qquad\square$

Corollary If $\mathbb{A}$ is a non-null closed subset of $\mathbb{R}^n$ and $\mathbb{B}$ is a non-null compact subset of $\mathbb{R}^n$, then there is a point α of $\mathbb{A}$ and a point β of $\mathbb{B}$ such that (5) holds.

Proof Let $\alpha_1 \in \mathbb{A}$ and $\beta_1 \in \mathbb{B}$. Let $\mathbb{I}$ be a compact interval containing $\mathbb{B}$, and $\mathbb{J}$ be a compact interval with the same centre and corresponding sides greater by $2|\alpha_1 - \beta_1|$. Then $\mathbb{A} \cap \mathbb{J}$ is compact, so there is a point α of $\mathbb{A} \cap \mathbb{J}$ (and therefore of $\mathbb{A}$) and a point β of $\mathbb{B}$ such that

$$|\xi - \eta| \geq |\alpha - \beta| \quad \text{whenever} \quad \xi \in \mathbb{A} \cap \mathbb{J} \text{ and } \eta \in \mathbb{B}. \quad (6)$$

Now let ξ be any point of $\mathbb{A}$ and η be any point of $\mathbb{B}$. If $\xi \notin \mathbb{J}$, then because ξ is outside $\mathbb{J}$ and η is in $\mathbb{I}$, $|\xi - \eta| \geq |\alpha_1 - \beta_1| \geq |\alpha - \beta|$ by (6). However, if $\xi \in \mathbb{J}$, then $|\xi - \eta| \geq |\alpha - \beta|$ by (6).

Exercises 16. Let f be in $\mathbb{R}$ into $\mathbb{R}$, dom f be compact, and f be continuous. Is

$$\{(x, f(x)) : x \in \text{dom} f\}$$

necessarily compact?

17. What happens in the corollary preceding these Exercises if $\mathbb{A}$ and $\mathbb{B}$ are both closed but neither is compact?

18. Is $\{(x, y) : 2x^2 + y^2 \leq 4\}$ compact? How about $\{(x, y) : 2x^2 + y^2 = 4\}$?

Problems 4. Is the converse to Exercise 16 true?

5. Generalize Exercise 16 and Problem 4 to higher-dimensional spaces.

Theorem If a one-to-one continuous function from $\mathbb{R}^n$ to $\mathbb{R}^m$ has a compact domain, its inverse is continuous.

Proof Let f be the function and g be its inverse. If $\mathbb{U}$ is any closed subset of $\mathbb{R}^m$ then $\mathbb{U} \cap (\text{dom} f)$ is compact. Then $f(\mathbb{U} \cap \text{dom} f)$ is compact and therefore closed. However, $f(\mathbb{U} \cap \text{dom} f) = g\check{}\mathbb{U}$. Thus $g\check{}\mathbb{U}$ is closed and therefore, by the corollary on p. 32, g is continuous. $\qquad \square$

Counter-example

If the domain of f is not compact, the theorem does not apply. For example, if $f(x) = (\cos x, \sin x)$ for $0 \leq x < 2\pi$, then f is a one-to-one continuous function in $\mathbb{R}$ into $\mathbb{R}^2$. Its inverse is not continuous at $(0, 1)$.

Exercises   19. Find a one-to-one continuous function in $\mathbb{R}^2$ into $\mathbb{R}$ whose inverse is not continuous. Also, find such a function in $\mathbb{R}^2$ into $\mathbb{R}^2$.

20. Find a one-to-one continuous function with closed domain whose inverse is not continuous.

Problems   6. Find a continuous function f with a continuous inverse and open domain, and a subset $\mathbb{S}$ of dom f for which bdy $\mathbb{S} \subseteq$ dom f but for which bdy $f(\mathbb{S})$ is not contained in rng f. (Note. Then bdy $f(\mathbb{S})$ cannot equal $f(\text{bdy } \mathbb{S})$.)

7(a). Find a continuous function f and a subset $\mathbb{S}$ of rng f for which $f^{\sim}(\text{int } \mathbb{S})$ contains a point not in int $f^{\sim}\mathbb{S}$. 7(b) The opposite to Exercise 7(a): int $f^{\sim}\mathbb{S}$ is to contain a point not in $f^{\sim}(\text{int } \mathbb{S})$.

8. Prove that if f and its inverse are continuous, dom f is open, bdy $\mathbb{S} \subseteq$ dom f, and bdy $f(\mathbb{S}) \subseteq$ rng f, then bdy $f(\mathbb{S}) = f(\text{bdy } \mathbb{S})$.

Uniform continuity

Let us look closely at the condition for a function f to be continuous on a set $\mathbb{S}$. The ϵ, δ criterion for continuity (p. 23) shows that f is continuous at α if and only if

given any positive number ϵ there is a positive number δ such that $|f(\xi) - f(\alpha)| < \epsilon$ whenever $|\xi - \alpha| < \delta$.

The criterion for f to be continuous on $\mathbb{S}$ at α is the same except for replacing the phrase 'whenever $|\xi - \alpha| < \delta$' by 'whenever $|\xi - \alpha| < \delta$ and $\xi \in \mathbb{S}$'. Therefore f is continuous on $\mathbb{S}$ if and only if

given any positive number ϵ and any element α of $\mathbb{S}$ there is a positive number δ such that $|f(\xi) - f(\alpha)| <$ (7) ϵ whenever $|\xi - \alpha| < \delta$ and $\xi \in \mathbb{S}$.

There is another criterion which may seem at first sight only trivially different:

given any positive number ϵ there is a positive number δ such that $|f(\xi) - f(\alpha)| < \epsilon$ whenever (8) $|\xi - \alpha| < \delta$, $\xi \in \mathbb{S}$ and $\alpha \in \mathbb{S}$.

The only substantial difference is the order in which α and δ are mentioned. However, this makes (8) much stronger than (7), and continuous functions do not by any means all have property (8). The reason is that in (7) α is given first and then a suitable δ has to be found. In (8), on the other hand, α is not given; a δ has to be found that is effective for *every* α in $\mathbb{S}$—it has to be effective 'uniformly in α'.

Examples   9. Let $\mathbb{S}$ be the interval $(0; 1)$ and let $f(x) = x^{-1}$ for every x in $\mathbb{S}$. Then f is continuous at every number in $\mathbb{S}$, so that (7) holds. How about (8)? Given $\epsilon > 0$ and $c \in \mathbb{S}$, δ has to be such that

$$|c^{-1} - x^{-1}| < \epsilon \qquad \text{whenever} \qquad |c - x| < \delta. \qquad (9)$$

It is the small values of ϵ that are vital, so let us suppose that $\epsilon < c^{-1}$. Then (9) is equivalent to

$$c/(1 + \epsilon c) < x < c/(1 - \epsilon c)$$
$$\text{whenever} \qquad c - \delta < x < c + \delta,$$

and so for (9) to hold δ must satisfy

$$c + \delta \leq c/(1 - \epsilon c) \qquad \text{and} \qquad c - \delta \geq c/(1 + \epsilon c),$$

i.e.
$$\delta \leq \epsilon c^2/(1 - \epsilon c) \qquad \text{and} \qquad \delta \leq \epsilon c^2/(1 + \epsilon c).$$

Therefore the largest δ that will satisfy (9) is

$$\epsilon c^2/(1 + \epsilon c).$$

Now suppose that ϵ is given and we have to find a δ that satisfies (9) for every c. It has to be smaller than (or at most equal to) $\epsilon c^2/(1 + \epsilon c)$ for every c in $(0; 1)$. However, we can make $\epsilon c^2/(1 + \epsilon c)$ as small as we like by taking c near enough to zero, so there is no positive δ that is small enough. Therefore (8) is not satisfied. We could also have seen that (8) does not hold from the graph of f. The graph of $y = x^{-1}$ between $x = 0$ and $x = 1$ is shown in Fig. 3.1. Imagine that a horizontal band of width 2ϵ can be slid vertically up and down. For (8) to hold there would have to be a number δ such that no matter where a vertical band of width 2δ is drawn, the portion of the graph inside this band can be covered by the movable horizontal band. It is clear from the shape of the graph that this is not so; by taking the vertical band close enough to the left-hand edge we can make it cut off a portion of the graph as tall as we like. A formal proof that (8) does not hold would go as follows. Given ϵ and δ, let c be the smaller of ϵ^{-1} and δ

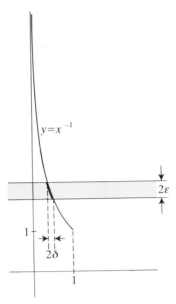

Fig. 3.1

and let $x = \frac{1}{2}c$. Then $|x - c| < \delta$ but

$$|f(x) - f(c)| = |x^{-1} - c^{-1}| = c^{-1} \geq \epsilon.$$

10. Let $\mathbb{S} = (0; 1)$ and $f(x) = x^2$ for every x. Let ϵ be any positive number, and let us investigate (8). We want to find a δ such that, for every c in $(0; 1)$,

$$|x^2 - c^2| < \epsilon \qquad \text{whenever} \qquad |x - c| < \delta \text{ and } x \in (0; 1).$$

Now $x^2 - c^2 = (x - c)(x + c)$ and so $|x^2 - c^2| < \delta \,|x + c|$ whenever $|x - c| < \delta$. In addition, $|x + c| < 2$ whenever $x \in (0; 1)$. Therefore

$$|x^2 - c^2| < 2\delta \qquad \text{whenever} \qquad |x - c| < \delta \text{ and } x \in (0; 1)$$

Therefore (8) holds in this case if we take δ to be $\frac{1}{2}\epsilon$.

We now give a formal definition.

Definition   If $\mathbb{S} \subseteq \text{dom} f$, then f is **uniformly continuous** on $\mathbb{S}$ if for each positive ϵ there is a positive δ such that

$$|f(\xi) - f(\eta)| < \epsilon \qquad \text{whenever} \qquad |\xi - \eta| < \delta$$
and both ξ and η belong to $\mathbb{S}$.

It follows at once that if f is uniformly continuous on $\mathbb{S}$ then f is continuous on $\mathbb{S}$.

The fundamental property of uniform continuity is that if f is continuous on a compact set $\mathbb{S}$ it is uniformly continuous on $\mathbb{S}$. We might try to prove this as follows. If ϵ is any positive number, then for each element ξ of $\mathbb{S}$ there is a positive δ_ξ such that $|f(\eta) - f(\xi)| < \epsilon$ whenever $|\eta - \xi| < \delta_\xi$. Let δ be the least of these δ_ξ. Then $|f(\eta) - f(\xi)| < \epsilon$ whenever $|\eta - \xi| < \delta$. Unfortunately this simple proof is not valid: there may be an infinite number of δ_ξ and there may not be a smallest one. We have to use the definition of compactness to select a finite number of the δ_ξ.

Theorem If f is continuous on a compact set $\mathbb{S}$, it is uniformly continuous on $\mathbb{S}$.

Proof Let ϵ be any positive number. For each ξ of $\mathbb{S}$ there is a positive δ_ξ such that

$$|f(\eta) - f(\xi)| < \tfrac{1}{2}\epsilon \quad \text{whenever} \quad |\eta - \xi| < \delta_\xi \text{ and } \eta \in \mathbb{S}.$$
$$(10)$$

The set of all open discs with centre ξ and radius $\tfrac{1}{2}\delta_\xi$ covers $\mathbb{S}$. Therefore a finite number of them will do so. Let δ be the radius of the smallest of this finite number of discs. If ξ and η belong to $\mathbb{S}$ and

$$|\xi - \eta| < \delta$$

then ξ will be in one of the finite number of open discs, say in the one with centre α. Then

$$|\xi - \alpha| < \tfrac{1}{2}\delta_\alpha.$$

However, $\delta \leq \tfrac{1}{2}\delta_\alpha$ and so

$$|\eta - \xi| < \tfrac{1}{2}\delta_\alpha.$$

Therefore

$$|\eta - \alpha| < \delta_\alpha.$$

Then, by (10),

$$|f(\eta) - f(\alpha)| < \tfrac{1}{2}\epsilon.$$

By (10) again

$$|f(\xi) - f(\alpha)| < \tfrac{1}{2}\epsilon$$

and so

$$|f(\xi) - f(\eta)| < \epsilon.$$

Exercises 21. Let f be the identity function on $\mathbb{R}$. We have seen that $1/f$ is not uniformly continuous on the interval $(0; 1)$ but f^2 is. Which of the following are uniformly continuous on this interval: $1/f^2$, $1/f^{1/2}$, $f^{1/2}$, f, and f^3?

22. Let f be the identity function on $\mathbb{R}$. Which, if any, of the functions $1/f^2$, $1/f$, $1/f^{1/2}$, $f^{1/2}$, f, and f^2 are uniformly continuous on $\{x : x > 0\}$?

23. Let $\mathbb{S} = \mathbb{R}$ and $f(x) = (x^2 + 1)^{-1}$ for every x of $\mathbb{R}$. Given a positive number ϵ, find a positive δ satisfying condition (8) or prove that no such δ exists.

24. Let f be the identity function on $\mathbb{R}^2$. Which of the functions $1/f^2$, f, and f^2 are uniformly continuous on $(0; 1) \times (0; 1)$?

25. $f(x, y) = xy$ for every (x, y) of $\mathbb{R}^2$. Is f uniformly continuous on $\mathbb{R}^2$?

26. Prove that if f is uniformly continuous on $\mathbb{R}^n$, then it is uniformly continuous on every subset of $\mathbb{R}^n$.

27. Prove that if f and g are uniformly continuous on $\mathbb{S}$, then so is $f + g$.

Problems 9. If $\mathbb{S}$ is an open disc, f is continuous at each element of $\mathbb{S}$, and $f(\mathbb{S})$ is bounded, must f be uniformly continuous on $\mathbb{S}$?

10. If $\mathbb{S}$ is an open disc and f is uniformly continuous on $\mathbb{S}$, need $f(\mathbb{S})$ be bounded?

11. If f is continuous everywhere in $\mathbb{R}^n$ and

$$\frac{|f(\xi) - f(\eta)|}{|\xi - \eta|} \le 100$$

for every pair of distinct elements ξ, η of $\mathbb{R}^n$, must f be uniformly continuous on $\mathbb{R}^n$?

12. If f is uniformly continuous on its domain, need there be a number m such that

$$\frac{|f(\xi) - f(\eta)|}{|\xi - \eta|} < m$$

for every pair of distinct elements ξ, η of the domain?

13. Let $\alpha \in \mathbb{R}^m$ and f be a function in $\mathbb{R}^n$ into $\mathbb{R}^m$. If f is uniformly continuous on $\mathbb{S}$, does it follow that $\alpha \cdot f$ is uniformly continuous on $\mathbb{S}$?

14. Is it true that f is uniformly continuous on $\mathbb{S}$ if and only if every component is?

15. $\mathbb{S}$ and $\mathbb{T}$ are compact sets and f is uniformly continuous on $\mathbb{S}$ and on $\mathbb{T}$. Is f necessarily uniformly continuous on $\mathbb{S} \cup \mathbb{T}$?

16. f and g are uniformly continuous on $\mathbb{S}$. Does it follow that fg is uniformly continuous on $\mathbb{S}$? What if f and g are bounded on $\mathbb{S}$?

17. If $\mathbb{S}$ is bounded and f is uniformly continuous on $\mathbb{S}$, does it follow that $f(\mathbb{S})$ is bounded?

18. f is uniformly continuous on a bounded set $\mathbb{S}$ and $\mathbb{S}^* = \mathbb{S} \cup \text{bdy } \mathbb{S}$. Is it always possible to find a function f^*, continuous on $\mathbb{S}^*$, such that $f^*(\xi) = f(\xi)$ whenever $\xi \in \mathbb{S}$?
19. If f is continuous on $\mathbb{R}^n$ must it be uniformly continuous on every bounded subset of $\mathbb{R}^n$?
20. If f is uniformly continuous on $\mathbb{S}$ and on $\mathbb{T}$ (not both compact), is f necessarily uniformly continuous on $\mathbb{S} \cup \mathbb{T}$?

Connected sets

If the set $\mathbb{S}$ consists of the open disc with centre $(0, 0)$ and radius 1 together with the points of the open disc with centre $(2, 2)$ and radius 1, then $\mathbb{S}$ clearly falls into two disconnected parts, as opposed to the points of one single open disc or the points of an interval. If $\mathbb{S}$ consists of the points of the open disc with centre $(1, 0)$ and radius 1 together with the points of the open disc with centre $(-1, 0)$ and radius 1 it is not so obvious that $\mathbb{S}$ is disconnected. If $\mathbb{S}$ consists of all points of the form $(x, \sin(1/x))$ together with all points of the form $(0, y)$, it is not at all clear whether $\mathbb{S}$ is connected, and we need a precise definition of a connected set if we are to come to a decision.

We base our definition of a connected set on the following thought experiment. Imagine that the set $\mathbb{S}$ is made of absorbent paper. Feed blue ink into $\mathbb{S}$ at one point. If a point turns blue, every point of $\mathbb{S}$ in its immediate neighbourhood turns blue as the ink spreads. This means that if α is a blue point there is a neighbourhood $\mathbb{U}_\alpha$ of α such that all of $\mathbb{S} \cap \mathbb{U}_\alpha$ turns blue. (Similarly if α does not turn blue.) If $\mathbb{S}$ is connected, it will all turn blue; if it is not, only part will.

We say that $\mathbb{S}$ is **split** into $\mathbb{A}$ and $\mathbb{B}$ if $\mathbb{A} \cup \mathbb{B} = \mathbb{S}$ and $\mathbb{A}$ and $\mathbb{B}$ have no common point.

Definition $\mathbb{S}$ is **connected** if it cannot be split into proper subsets $\mathbb{A}$ and $\mathbb{B}$ with the property that each point α of $\mathbb{A}$ has a neighbourhood $\mathbb{U}_\alpha$ such that

$$\mathbb{S} \cap \mathbb{U}_\alpha \subseteq \mathbb{A}$$

and each point β of $\mathbb{B}$ has a neighbourhood $\mathbb{U}_\beta$ such that

$$\mathbb{S} \cap \mathbb{U}_\beta \subseteq \mathbb{B}. \qquad \square$$

Recall that $\mathbb{P}$ is open if each point α of $\mathbb{P}$ has a neighbourhood $\mathbb{U}_\alpha$ such that $\mathbb{U}_\alpha \subseteq \mathbb{P}$. It is natural to make the following definition.

Definition $\mathbb{P}$ is **open in** $\mathbb{Q}$ if each point α of $\mathbb{P}$ has a neighbourhood $\mathbb{U}_\alpha$ such that

$$\mathbb{Q} \cap \mathbb{U}_\alpha \subseteq \mathbb{P}.$$

Then the $\mathbb{A}$ and $\mathbb{B}$ in the definition of a connected set are open in $\mathbb{S}$, and we can say that $\mathbb{S}$ is connected if it cannot be split into proper subsets open in $\mathbb{S}$.   □

The following theorem allows us to display a variety of connected sets.

Theorem   Let $\mathbb{S}$ be a set with the property that whenever α and β are points of $\mathbb{S}$ they can be joined by a sequence of line segments lying entirely in $\mathbb{S}$. Then $\mathbb{S}$ is connected.

Proof   Suppose that $\mathbb{S}$ can be split into proper subsets $\mathbb{A}$ and $\mathbb{B}$ open in $\mathbb{S}$. Let α be a point of $\mathbb{A}$ and β a point of $\mathbb{B}$. Then α and β can be joined by a sequence of line segments, and so there is a line segment with one end lying in $\mathbb{A}$ and the other in $\mathbb{B}$. Call this line segment $\mathbb{L}$. Each point of $\mathbb{L}$ has a neighbourhood $\mathbb{U}$ such that

$$\mathbb{U} \cap \mathbb{L}$$

is either contained in $\mathbb{A}$ or contained in $\mathbb{B}$. The set of all such $\mathbb{U}$ form an open covering of $\mathbb{L}$ and so a finite number of them suffice to cover $\mathbb{L}$ because $\mathbb{L}$, being closed and bounded, is compact. The corresponding $\mathbb{U} \cap \mathbb{L}$ will also cover $\mathbb{L}$. However, any two of these sets which overlap must either both be in $\mathbb{A}$ or both be in $\mathbb{B}$. This is clearly impossible if one endpoint is in $\mathbb{A}$ and the other in $\mathbb{B}$. Therefore $\mathbb{S}$ must be connected.   □

It follows from this theorem that $\mathbb{R}^n$ is connected, open discs are connected, and intervals are connected. The theorem we have just proved has a partial converse.

Theorem   If $\mathbb{S}$ is an open connected set and α and β are any two points of $\mathbb{S}$, then α and β can be joined by a sequence of line segments lying entirely in $\mathbb{S}$.

Proof   Let $\mathbb{A}$ be the set of all points of $\mathbb{S}$ that can be joined to α by a sequence of line segments in $\mathbb{S}$ and let $\mathbb{B}$ be $\mathbb{S}\backslash\mathbb{A}$. Let τ be any point of $\mathbb{A}$. Because $\mathbb{S}$ is open there is a neighbourhood $\mathbb{U}_\tau$ of τ contained in $\mathbb{S}$. Any point in $\mathbb{U}_\tau$ can be joined to τ by a line segment lying in $\mathbb{U}_\tau$ and so can be joined to α by a sequence of line segments lying in $\mathbb{S}$. Then $\mathbb{U}_\tau \subseteq \mathbb{A}$, i.e. $\mathbb{S} \cap \mathbb{U}_\tau \subseteq \mathbb{A}$. Now let τ be any point of $\mathbb{B}$. Again, there is a neighbourhood $\mathbb{U}_\tau$ of τ contained in $\mathbb{S}$, every point of which can be joined to τ by a line segment in $\mathbb{S}$ and therefore cannot be joined to α by a sequence of line segments (because if it could, then τ could). Therefore $\mathbb{U}_\tau \subseteq \mathbb{B}$, i.e. $\mathbb{S} \cap \mathbb{U}_\tau \subseteq \mathbb{B}$. Then, because $\mathbb{S}$ is connected, $\mathbb{A}$ and $\mathbb{B}$ cannot be proper subsets. $\mathbb{A}$ contains α and so $\mathbb{A} = \mathbb{S}$. Thus β can be joined to α. □

There are two interesting results involving connected sets and continuous functions.

Theorem   If f is a function from $\mathbb{R}^n$ to $\mathbb{R}^m$, if its domain $\mathbb{S}$ is connected, and if f is continuous on $\mathbb{S}$, then $f(\mathbb{S})$ is connected.

Proof   If not, let $f(\mathbb{S})$ split into proper subsets $\mathbb{A}$ and $\mathbb{B}$ open in $f(\mathbb{S})$. Because $\mathbb{A}$ and $\mathbb{B}$ have no point in common, the inverse images $f^{\sim}\mathbb{A}$ and $f^{\sim}\mathbb{B}$ have no point in common. Because $\mathbb{A}$ and $\mathbb{B}$ are non-empty subsets of $f(\mathbb{S})$, $f^{\sim}\mathbb{A}$ and $f^{\sim}\mathbb{B}$ are non-empty subsets of $\mathbb{S}$. Because $\mathbb{A} \cup \mathbb{B} = f(\mathbb{S})$, $(f^{\sim}\mathbb{A}) \cup (f^{\sim}\mathbb{B}) = \mathbb{S}$. Thus $\mathbb{S}$ splits into proper subsets $f^{\sim}\mathbb{A}$ and $f^{\sim}\mathbb{B}$.

If $\alpha \in f^{\sim}\mathbb{A}$, $f(\alpha) \in \mathbb{A}$ and so, because $\mathbb{A}$ is open in $f(\mathbb{S})$, there is a neighbourhood $\mathbb{U}$ of $f(\alpha)$ such that

$$f(\mathbb{S}) \cap \mathbb{U} \subseteq \mathbb{A}.$$

Then if $f(\zeta) \in \mathbb{U}$ it follows that $f(\zeta) \in \mathbb{A}$; therefore

$$f^{\sim}\mathbb{U} \subseteq f^{\sim}\mathbb{A}.$$

Because $f(\alpha) \in \mathbb{U}$ and $\mathbb{U}$ is open there is a neighbourhood $\mathbb{V}$ of α such that

$$\mathbb{V} \cap \mathbb{S} \subseteq f^{\sim}\mathbb{U}$$
$$\subseteq f^{\sim}\mathbb{A}.$$

Thus $f^{\sim}\mathbb{A}$ is open in $\mathbb{S}$. Similarly, $f^{\sim}\mathbb{B}$ is open in $\mathbb{S}$.

However, this is not possible, because $\mathbb{S}$ is connected. Thus $f(\mathbb{S})$ cannot be split as described and must be connected.   □

Intermediate-value theorem

If α and β are points of a connected set $\mathbb{S}$, if f is continuous on $\mathbb{S}$ and is into $\mathbb{R}$, and if κ lies between $f(\alpha)$ and $f(\beta)$, then there is a point γ in $\mathbb{S}$ such that $f(\gamma) = \kappa$.

Proof   If not, $\mathbb{S}$ splits into proper subsets $\{\tau \in \mathbb{S} : f(\tau) > \kappa\}$ and $\{\tau \in \mathbb{S} : f(\tau) < \kappa\}$ open in $\mathbb{S}$.   □

Exercises   28. Let $\mathbb{S}$ consist of the points of the open disc with centre $(0, 1)$ and radius 1 together with the points of the open disc with centre $(0, -1)$ and radius 1. Prove that $\mathbb{S}$ is not connected.

29. Prove that $\{(x, y) : 1 \leq x^2 + y^2 \leq 2\}$ is connected.

30. Prove that $\mathbb{S}$ is connected if and only if whenever $\mathbb{S}$ is split into two proper subsets one of them contains a boundary point of the other.

31. Let $\mathbb{S}$ consist of all points of the form $(x, \sin(1/x))$ for all positive x together with all points of the form $(0, x)$. Is $\mathbb{S}$ connected?

32. Let $\mathbb{A}$ and $\mathbb{B}$ be connected subsets of $\mathbb{R}^n$ with no points in common but with the property that $\inf\{|\alpha - \beta| : \alpha \in \mathbb{A}$ and $\beta \in \mathbb{B}\} = 0$. Must $\mathbb{A} \cup \mathbb{B}$ be connected?

33. Show that $\mathbb{S}$ is disconnected if and only if there are open sets $\mathbb{P}$ and $\mathbb{Q}$ such that $\mathbb{S}$ splits into proper subsets $\mathbb{S} \cap \mathbb{P}$ and $\mathbb{S} \cap \mathbb{Q}$.

34. $\mathbb{S}$ is connected and $\mathbb{T}$ consists of $\mathbb{S}$ together with some or all of its boundary points. Prove that $\mathbb{T}$ is connected.

35. Prove that every connected subset of $\mathbb{R}$ is an interval, but that if $n > 1$ there are connected subsets of $\mathbb{R}^n$ that are not intervals.

Problems   21. If $\mathbb{S}$ is connected, is its interior necessarily connected?

22. If $\mathbb{A}$ and $\mathbb{B}$ are connected, does it follow that their cartesian product is connected?

23. If f is continuous and $\mathbb{S}$ is a connected subset of $\text{rng} f$, does it follow that the inverse image of $\mathbb{S}$ under f is connected?

24. If f is one-to-one and continuous and $\mathbb{S}$ is a connected subset of $\text{rng} f$, does it follow that the inverse image of $\mathbb{S}$ under f is connected?

Limits

The idea of 'limit' in many dimensions is the same as in one dimension: λ is the limit of the function f at α if we can keep

$f(\xi)$ close to λ by keeping ξ close to α. This can be expressed quite neatly in terms of neighbourhoods.

Definition If f is in $\mathbb{R}^n$ into $\mathbb{R}^m$, if $\alpha \in \mathbb{R}^n$, if $\lambda \in \mathbb{R}^m$, and if, for each neighbourhood $\mathbb{N}$ of λ, there is a neighbourhood $\mathbb{U}$ of α such that

$$f(\xi) \in \mathbb{N} \quad \text{whenever} \quad \xi \in \mathbb{U} \quad \text{but} \quad \xi \neq \alpha,$$

then λ is a **limit** of f at α. □

Theorem (uniqueness of the limit)
If λ and μ are limits of f at α, then $\lambda = \mu$.

Proof If λ were not equal to μ we could find neighbourhoods $\mathbb{L}$ and $\mathbb{M}$ of λ and μ respectively with no point in common. Then there would be neighbourhoods $\mathbb{U}$ and $\mathbb{V}$ of α such that

$$f(\xi) \in \mathbb{L} \quad \text{whenever} \quad \xi \in \mathbb{U} \quad \text{but} \quad \xi \neq \alpha$$
and
$$f(\xi) \in \mathbb{M} \quad \text{whenever} \quad \xi \in \mathbb{V} \quad \text{but} \quad \xi \neq \alpha.$$

$\mathbb{U} \cap \mathbb{V}$ would be a neighbourhood of α and would contain a point η other than α. Then $f(\eta)$ would lie in both $\mathbb{L}$ and $\mathbb{M}$, which is impossible. Thus f cannot have more than one limit at α. If it has one, we denote the limit by

$$\lim_\alpha f$$
or
$$\lim_{\xi \to \alpha} f(\xi)$$

provided that the letter ξ is not already in use in the context. □

There is a fundamental connection between limits and continuity.

Theorem f is continuous at α if and only if $\lim_\alpha f = f(\alpha)$.

Proof First, suppose that f is continuous at α. Let $\mathbb{N}$ be any neighbourhood of $f(\alpha)$. By definition of continuity, there is a neighbourhood $\mathbb{U}$ of α such that $f(\xi) \in \mathbb{N}$ whenever $\xi \in \mathbb{U}$. Consequently, $f(\alpha)$ is a limit of f at α. Conversely, suppose that $\lim_\alpha f = f(\alpha)$. Let $\mathbb{N}$ be any neighbourhood of $f(\alpha)$. By definition of limit, there is a neighbourhood $\mathbb{U}$ of α such that $f(\xi) \in \mathbb{N}$ whenever $\xi \in \mathbb{U}$ but $\xi \neq \alpha$. However, if $\xi = \alpha$ then

$f(\xi) = f(\alpha) \in \mathbb{N}$. Thus $f(\xi) \in \mathbb{N}$ *whenever* $\xi \in \mathbb{U}$. Therefore f is continuous at α. □

This means that for each of the fundamental theorems on continuity there is a corresponding theorem on limits.

Theorems
1. If α is interior to dom f and f is an identity function or is constant, then $\lim_\alpha f$ exists and equals $f(\alpha)$.
2. $\lim_\alpha f = \lambda$ if and only if $\lim_\alpha f_i = \lambda_i$ for each i.
3. (The ϵ, δ criterion.) $\lim_\alpha f = \lambda$ if and only if for each positive ϵ there is a positive δ such that $|f(\xi) - \lambda| < \epsilon$ whenever $0 < |\xi - \alpha| < \epsilon$.
4. (Limit of a sum.) If f, g, h, k, etc. are functions in $\mathbb{R}^n$ into $\mathbb{R}^m$ that have limits at α, and if $\beta \in \mathbb{R}^m$, then

$$\lim_\alpha (f + g) = \lim_\alpha f + \lim_\alpha g,$$

$$\lim_\alpha (f + g + \beta) = \lim_\alpha f + \lim_\alpha g + \beta,$$

$$\lim_\alpha (f + g + h + k) = \lim_\alpha f + \lim_\alpha g + \lim_\alpha h + \lim_\alpha k, \text{ etc.}$$

5. If $\lim_\alpha f$ exists and $c \in \mathbb{R}$, then $\lim_\alpha (cf) = c \lim_\alpha f$.
6. If f is in $\mathbb{R}^n$ into $\mathbb{R}^m$ and $\gamma \in \mathbb{R}^m$, then $\lim_\alpha (\gamma \cdot f) = \gamma \cdot \lim_\alpha f$.
7. (Limit of a product.) If f and g are in $\mathbb{R}^n$ into $\mathbb{R}^m$ and $\lim_\alpha f$ and $\lim_\alpha g$ exist, then $\lim_\alpha (f \cdot g) = \lim_\alpha f \cdot \lim_\alpha g$.
8. If F is in $\mathbb{R}^n$ into $\mathbb{R}$ and f is in $\mathbb{R}^n$ into $\mathbb{R}^m$, and $\lim_\alpha F$ and $\lim_\alpha f$ exist, then $\lim_\alpha (Ff) = \lim_\alpha F \lim_\alpha f$.
9. If F is in $\mathbb{R}^n$ into $\mathbb{R}$ and $\lim_\alpha F$ exists and is not zero, then $\lim_\alpha (1/F) = 1/\lim_\alpha F$. □

(Note. We do not have a 'limit of a composite' theorem. See, however, Problem 26.)

These theorems can be proved easily from the corresponding continuity theorems. Here is one example.

Theorem
If f is a function in $\mathbb{R}^n$ into $\mathbb{R}^m$ and $\lim_\alpha f = \lambda$, then $\lim_\alpha f_i = \lambda_i$ for each i.

Proof
Define a function g by setting

$$\begin{cases} g_i(\alpha) = \lambda_i \\ g_i(\xi) = f_i(\xi) & \text{whenever} \quad \xi \in \text{dom} f \quad \text{but} \quad \xi \neq \alpha \end{cases}$$

for each i. Then g is continuous at α. Therefore so is each g_i. Therefore $\lim_\alpha g_i = \lambda_i$. But $\lim_\alpha f_i = \lim_\alpha g_i = \lambda_i$. □

Exercises 36. In each of the following cases investigate the limit of f at $(0, 0)$:
(a) $f(x, y) = x^2/(x^2 + y^2)$ for every (x, y) of $\mathbb{R}^2$, except $(0, 0)$;
(b) $f(x, y) = xy/(x^2 + y^2)$ for every (x, y) of $\mathbb{R}^2$, except $(0, 0)$;
(c) $f(x, y) = x^3/(x^2 + y^2)$ for every (x, y) of $\mathbb{R}^2$, except $(0, 0)$;
(d) $f(x, y) = (xy)^{4/3}/(x^2 + y^2)$ for every (x, y) of $\mathbb{R}^2$ except $(0, 0)$.

37. In each of the following cases investigate $\lim_{(0,0)} f$, $\lim_0 f(\cdot, 0)$, $\lim_0 f(0, \cdot)$, $\lim_{x \to 0}\{\lim_0 f(x, \cdot)\}$, and $\lim_{y \to 0}\{\lim_0 f(\cdot, y)\}$:

(a) $f(x, y) = \begin{cases} xy/|xy| & \text{if } xy \neq 0 \\ 0 & \text{if } xy = 0; \end{cases}$

(b) $f(x, y) = xy/(x^2 + y^2)$ for every (x, y) of $\mathbb{R}^2$, except $(0, 0)$;
(c) $f(x, y) = (x^2 - y^2)/(x^2 + y^2)$ for every (x, y) of $\mathbb{R}^2$, except $(0, 0)$.†

38. Is it true that $\lim_\alpha f = \lambda$ if and only if, for each i,

$$\lim_{\alpha_i} f(\alpha_1, \alpha_2, \ldots, \alpha_{i-1}, \cdot, \alpha_{i+1}, \ldots, \alpha_n) = \lambda?$$

Problems 25. Given that $\lim_\alpha f = \beta$ and that f has an inverse f^-, what can we say about $\lim_\beta f^-$?

26. If f is in $\mathbb{R}^n$ into $\mathbb{R}^m$, h is in $\mathbb{R}^m$ into $\mathbb{R}^p$, $\lim_\alpha f = \beta$, and $\lim_\beta h = \gamma$, what can we deduce about $\lim_\alpha h \circ f$? (Must it exist? If so, can we say what it is? If not, are there any simple extra conditions which will ensure that it exists?)

27. Take any two of the limits mentioned in Exercise 37, say $\lim_{(0,0)} f$ and $\lim_0 f(\cdot, 0)$. If one exists, does the other necessarily exist? If they both exist, does it follow that they are equal? Answer the same questions for every other pair of these limits.

The double limit and limits along lines

Let $\lim_{(0,0)} f = \lambda$ and let f_α be defined by

$$f_\alpha(x) = f(x \cos \alpha, x \sin \alpha) \qquad \text{for each } x.$$

Diagrammatically, the values of f_α are the values of f at points on the line through $(0, 0)$ with inclination α to the x axis. Thus we should expect that $\lim_0 f_\alpha = \lambda$, and that this should be so for every α. This is in fact true, and is easily proved.

A converse of this statement would be

if $\lim_0 f_\alpha = \lambda$ for every α,
then $\lim_{(0,0)} f = \lambda$.

† See p. 6 for the definition of $f(\cdot, 0)$ etc.

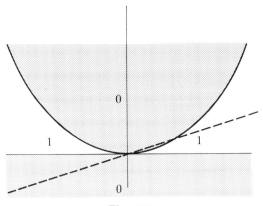

Fig. 3.2

However, this is false, as we see by taking f as follows:

$$f(x, y) = \begin{cases} 1 & \text{if } 0 < y < x^2 \\ 0 & \text{otherwise.} \end{cases}$$

Diagrammatically, the value of f is 1 between the parabola $y = x^2$ and the x axis, and zero elsewhere (Fig. 3.2). Then $\lim_0 f_\alpha = 0$ for every α. However, $\lim_{(0,0)} f$ does not exist, because every neighbourhood of $(0, 0)$ contains both points where the value of f is 1 and points where its value is zero.

The logic involved here is worth examining. One reason for supposing that the converse is true might be put as follows: for *any* α, the values of f_α are arbitrarily close to λ at points near enough to $(0, 0)$, and any value of f is a value of f_α for some suitable α. Therefore values of f are arbitrarily close to λ at points near enough to $(0, 0)$. However, an analysis of what is meant by 'arbitrarily close' shows that this argument is not sound. Let ϵ be any positive number. It is true that, given α, there is a positive δ such that

$$\begin{aligned} &f_\alpha(x) \text{ is within } \epsilon \text{ of } \lambda \text{ whenever} \\ &x \text{ is within } \delta \text{ of } 0 \text{ but } x \neq 0. \end{aligned} \tag{11}$$

This is true for *each* α, but δ might depend on α; we cannot say that there is any one δ such that (11) holds for *every* α. We cannot save the argument by taking the least of the various δs we obtain—there may be an infinite number of them, and there may not be a least one. Nor can we take their greatest lower bound, because this might be zero.

4
Differentiation

If f is a function in $\mathbb{R}^2$ into $\mathbb{R}$ we can define two derivatives, D_1f and D_2f, as follows:

$$D_1f(x, y) = f(\cdot, y)'(x)$$
$$D_2f(x, y) = f(x, \cdot)'(y)$$

where $f(\cdot, y)$ and $f(x, \cdot)$ are the functions defined on p. 6. These two derivatives have obvious interpretations: the derivative of f with the second argument held constant at y and the derivative with the first argument held constant at x respectively.

Exercise 1. Let $f(x, y) = x^3y^4$ for every x and y. Find $D_1f(x, y)$ and $D_2f(x, y)$.

In general, if f is in $\mathbb{R}^n$ into $\mathbb{R}$ we make the following definition.

Definition $D_if(x_1, \ldots, x_n)$ is the value of $f(x_1 \ldots x_{i-1}, \cdot, x_{i+1} \ldots x_n)'$ at x_i. The D_if are called the **partial derivatives** of f. □

Exercises 2. If $f(x, y) = x^2 - y^2$ for every (x, y) in $\mathbb{R}^2$, what are D_1f, D_2f, $D_1f(a, b)$, $D_2f(x, x)$, and $D_1f(y, x)$?
3. If $f(x, y, z) = xyz$ for every (x, y, z) in $\mathbb{R}^3$, what are $D_1f(1, 2, 3)$, $D_2f(1, 2, 3)$, and $D_3f(1, 2, 3)$?
4. If $f(0, 0) = 0$ and $f(x, y) = (x^2 - y^2)/(x^2 + y^2)$ whenever $(x, y) \neq (0, 0)$, do $D_1f(0, 0)$ and $D_2f(0, 0)$ exist?
5. If $f(x, y) = \arctan(y/x)$ whenever $x \neq 0$, what are D_1f and D_2f?
6. If f is in $\mathbb{R}^3$ into $\mathbb{R}$, express $D_2f(x, y, z)$ as a limit.
7. If g is in $\mathbb{R}^2$ into $\mathbb{R}$ and $f(x, y) = g(x, y) - g(y, x)$, prove that $D_1f(x, y) = -D_2f(y, x)$.

Problems 1. If $D_1f(a, b)$ and $D_2f(a, b)$ exist, must $f(a, b)$ be defined? If so, must f be continuous at (a, b)?

2. If $D_1 f$ and $D_2 f$ are continuous at (a, b), must f be continuous at (a, b)?
3. $f(x, y) = x^3/(x^2 + y^2)$ whenever $(x, y) \neq (0, 0)$, and $f(0, 0) = 0$. Are $D_1 f$ and $D_2 f$ continuous at $(0, 0)$?

Theorem Let

u be in $\mathbb{R}$ into $\mathbb{R}^2$ and u_1 and u_2 be differentiable at x

and

f be in $\mathbb{R}^2$ into $\mathbb{R}$ and $D_1 f$ and $D_2 f$ be continuous at $u(x)$.

Then

$$(f \circ u)'(x) = D_1 f(u(x)) \cdot u_1'(x) + D_2 f(u(x)) \cdot u_2'(x). \quad (1)$$

Proof Because $D_1 f$ and $D_2 f$ are continuous at $u(x)$, there is a neighbourhood $\mathbb{N}$ of $u(x)$ contained in their domains. Because u_1 and u_2 are continuous at x, there is a neighbourhood $\mathbb{M}$ of x such that

$$u(y) \in \mathbb{N} \qquad \text{whenever} \qquad y \in \mathbb{M}.$$

Let δ be any number for which

$$x + \delta \in \mathbb{M}.$$

Set

$$g = f(\cdot, u_2(x + \delta)) \qquad \text{and} \qquad h = f(u_1(x), \cdot).$$

Then

$$g'(z) = D_1 f(z, u_2(x + \delta)).$$

If y lies between $u_1(x)$ and $u_1(x + \delta)$, then

$$(y, u_2(x + \delta)) \in \mathbb{N} \subseteq \text{dom } D_1 f$$

and so $y \in \text{dom } g'$. Therefore we can apply the mean-value theorem to g, and so

$$g(u_1(x + \delta)) - g(u_1(x)) = \{u_1(x + \delta) - u_1(x)\} \cdot g'(\theta) \quad (2)$$

for some θ between $u_1(x)$ and $u_1(x + \delta)$. Similarly,

$$h(u_2(x + \delta)) - h(u_2(x)) = \{u_2(x + \delta) - u_2(x)\} \cdot h'(\phi) \quad (3)$$

for some ϕ between $u_2(x)$ and $u_2(x + \delta)$. If we add equations (2) and (3) using the three facts

$$h(u_2(x + \delta)) = g(u_1(x))$$
$$g(u_1(x + \delta)) = f(u(x + \delta))$$
$$h(u_2(x)) = f(u(x))$$

we obtain

$$(f \circ u)(x + \delta) - (f \circ u)(x) = \{u_1(x + \delta) - u_1(x)\} \cdot g'(\theta)$$
$$+ \{u_2(x+\delta) - u_2(x)\} \cdot h'(\phi). \quad (4)$$

Because $D_1 f$ is continuous at $u(x)$, u_1 is continuous at x, and θ is between $u_1(x)$ and $u_1(x + \delta)$, it follows that

$$\lim_{\delta \to 0} g'(\theta) = D_1 f(u(x)).$$

Similarly,

$$\lim_{\delta \to 0} h'(\phi) = D_2 f(u(x)).$$

Therefore, if we divide both sides of (4) by δ and take the limit as $\delta \to 0$, we obtain (1). $\qquad \square$

This theorem shows that the continuity of the partial derivatives is of some interest; several later theorems will also show this. We make the following definitions.

Definition A function is **continuously differentiable** at α if each of its partial derivatives is continuous at α. A function is **continuously differentiable** if it is continuously differentiable at each point of its domain. $\qquad \square$

Notice that a function cannot be continuously differentiable at α unless α is *interior* to the domains of its partial derivatives, and a function cannot be continuously differentiable unless its domain is open.

Note Some authors use the term 'of class $C^{(1)}$' to mean 'continuously differentiable'. In general, a function is of class $C^{(r)}$ if its rth-order partial derivatives (see p. 53) are continuous at each point of its domain; it is of class $C^{(\infty)}$ if all higher-order partial derivatives are continuous at each point of its domain.

Corollary 1

If f is in $\mathbb{R}^2$ into $\mathbb{R}$, if u is in $\mathbb{R}$ into dom f, and if u and f are continuously differentiable, then

$$[f \circ u]' = D_1 f \circ u \cdot u_1' + D_2 f \circ u \cdot u_2'. \qquad \square$$

Naturally, both the theorem and the corollary can be generalized to the case where f is in $\mathbb{R}^n$. The corollary will then become as follows.

Corollary 2

If f is in $\mathbb{R}^n$ into $\mathbb{R}$, if u is in $\mathbb{R}$ into dom f, and if u and f are continuously differentiable, then

$$[f \circ u]' = \sum D_i f \circ u \cdot u_i'. \qquad \square$$

Note

The conclusion of the theorem holds under somewhat weaker conditions than the ones we used: it holds even if $D_2 f$ is not continuous at $u(x)$, as long as $D_2 f(u(x))$ exists. (To prove this we apply the definition of differentiability to h instead of the mean-value theorem.)

Similarly, if f is in $\mathbb{R}^n$, we need only $n - 1$ of the n partial derivatives to be continuous. In particular, if $n = 1$, the one and only (partial) derivative of f need not be continuous. That is why there is no mention of *continuous* differentiability in the familiar one-dimensional theorem.

Exercises

8. Let (x, y) be the identity function on $\mathbb{R}^2$ and t the identity function on $\mathbb{R}$. If $w = f(u, v)$ where $f = x^2 - xy + 2y$, $u = t \sin t$, and $v = (1 + t^2)^{1/2}$, find $w'(0)$.

9. Write out corollary 2 in the case $n = 3$. (u, v, w may be used in place of u_1, u_2, u_3 if preferred.)

10. Let (x, y, z) be the identity function on $\mathbb{R}^3$ and t the identity function on $\mathbb{R}$. If $q = f(u, v, w)$ where $f = x^3 + yz$, $u = \sin$, $v = (1 + t^2)^{1/2}$, and $w = t^2$, find $q'(0)$.

11. Let $q = f(\sin, \cos)$, where f is a continuously differentiable function in $\mathbb{R}^2$ into $\mathbb{R}$. Find a formula for q' in terms of the derivatives of f.

12. If f, u, and v are continuously differentiable functions in $\mathbb{R}^2$ into $\mathbb{R}$, find both partial derivatives of $f(u, v)$ in terms of the partial derivatives of f, u, and v.

Corollary 2 can be generalized further. (The reader who has solved Exercise 12 will probably guess what we are about to say.) Let us first suppose that instead of being in $\mathbb{R}$, u is in $\mathbb{R}^2$. Let (a, b) be any element of dom u, and $v = f \circ u$. Then

$$v(\cdot, b) = f \circ (u(\cdot, b))$$

and so, by corollary 1,

$$v(\cdot, b)' = \sum D_i f \circ (u(\cdot, b)) \cdot u_i(\cdot, b)' \qquad (5)$$

However, $v(\cdot, b)'(a) = D_1 v(a, b)$ and $u_i(\cdot, b)' = D_1 u_i(a, b)$.

Therefore, by (5) evaluated at a,

$$D_1 v(a, b) = \sum D_i f(u(a, b)) \cdot D_1 u_i(a, b).$$

Thus

$$D_1 v = \sum D_i f \circ u \cdot D_1 u_i.$$

Clearly a similar argument applies to D_2; the argument generalizes in the obvious way to the case where u is in $\mathbb{R}^m$. Thus we have the following corollary.

Corollary 3

If f is in $\mathbb{R}^n$ into $\mathbb{R}$, if u is in $\mathbb{R}^m$ into dom f, and if u and f are continuously differentiable, then

$$D_j(f \circ u) = \sum_i D_i f \circ u \cdot D_j u_i$$

for each j from 1 to m. ☐

We can make one further generalization very easily: if f is into $\mathbb{R}^p$ instead of into $\mathbb{R}$, we apply corollary 3 to each component of f. This gives our main corollary, which includes the others as special cases.

Main corollary

If f is in $\mathbb{R}^n$ into $\mathbb{R}^p$, if u is in $\mathbb{R}^m$ into dom f, and if u and f are continuously differentiable, it follows that

if $v = f \circ u$, then

$$D_j v_k = \sum_i D_i f_k \circ u \cdot D_j u_i$$

for each j from 1 to m and each k from 1 to p. ☐

Exercises 13. Write out the main corollary in the following cases: $m = 3$, $n = 1$, $p = 1$; $m = 1$, $n = 3$, $p = 1$; $m = n = p = 2$.

14. If $f = (x^2 \sin z, \cos(yz))$, where (x, y, z) is the identity on $\mathbb{R}^3$, find the six functions $D_i f_j$.

15. u, v, r, and s are continuously differentiable functions in $\mathbb{R}$ into $\mathbb{R}$ for which

$$u = r \cos s \qquad \text{and} \qquad v = r \sin s.$$

f and g are continuously differentiable functions in $\mathbb{R}^2$ into $\mathbb{R}$ such that

$$g(r, s) = f(u, v).$$

Express the partial derivatives of g in terms of those of f.

16. In Exercise 15 we say informally 'the relations $u = r \cos s$ and $v = r \sin s$ transform f into g'. Into what do the relations $r = x + y$ and $s = x - y$ transform the equation

$$D_1 f(x, y) + D_2 f(x, y) = 0?$$

Find as general a function f as you can for which

$$D_1 f + D_2 f = 0.$$

Higher-order derivatives

If f is a function in $\mathbb{R}^n$ into $\mathbb{R}$, then so is each of its partial derivatives $D_i f$. Each of these partial derivatives will itself have partial derivatives. A partial derivative of a partial derivative of f is called a **second-order** partial derivative of f. A partial derivative of a second-order partial derivative is called a **third-order** partial derivative, and so on.

Example   1. Let $f(x, y) = x^2 y^3$ for every (x, y) of $\mathbb{R}^2$. Then

$$D_1 f(x, y) = 2xy^3$$

from which we obtain

$$D_1 D_1 f(x, y) = 2y^3 \quad \text{and} \quad D_2 D_1 f(x, y) = 6xy^2.$$

These formulae give two of the second-order partial derivatives of f. From

$$D_2 f(x, y) = 3x^2 y^2$$

we obtain the other two:

$$D_1 D_2 f(x, y) = 6xy^2 \quad \text{and} \quad D_2 D_2 f(x, y) = 6x^2 y.$$

The third-order partial derivatives include $D_1 D_2 D_2 f$. We have

$$D_1 D_2 D_2 f(x, y) = 12xy.$$

(Note. We often abbreviate $D_1 D_1 f$ to $D_1^2 f$ etc.)

Mixed second-order derivatives

In the above example $D_2 D_1 f = D_1 D_2 f$. A little experimentation with various functions f shows that this usually happens. In fact, we can prove that these two 'mixed second-order derivatives' are equal wherever they are continuous.

Theorem Let f be a function in $\mathbb{R}^2$ into $\mathbb{R}$. If D_1D_2f and D_2D_1f are continuous at (a, b), then

$$D_1D_2f(a, b) = D_2D_1f(a, b).$$

Proof If not, let

$$|D_1D_2f(a, b) - D_2D_1f(a, b)| = c.$$

Then there is a neighbourhood $\mathbb{M}$ of a and a neighbourhood $\mathbb{N}$ of b such that

$$|D_1D_2f(u, v) - D_1D_2f(a, b)| < \tfrac{1}{2}c$$

and

$$|D_2D_1f(u, v) - D_2D_1f(a, b)| < \tfrac{1}{2}c$$

whenever $u \in \mathbb{M}$ and $v \in \mathbb{N}$. Let h and k be non-zero and such that $a + h$ lies in $\mathbb{M}$ and $b + k$ in $\mathbb{N}$, and let g be defined by

$$g(x) = f(x, b + k) - f(x, b) \tag{6}$$

for every x in $\mathbb{M}$. Then

$$g'(x) = D_1f(x, b + k) - D_1f(x, b)$$

and so, by the mean-value theorem,

$$g(a + h) - g(a) = h \cdot \{D_1f(x_1, b + k) - D_1f(x_1, b)\}$$

for some x_1 between a and $a + h$ and consequently in $\mathbb{M}$. By the mean-value theorem again,

$$D_1f(x_1, b + k) - D_1f(x_1, b) = kD_2D_1f(x_1, y_1)$$

for some y_1 between b and $b + k$ and consequently in $\mathbb{N}$. Then

$$f(a + h, b + k) - f(a + h, b) - f(a, b + k) + f(a, b)$$
$$= g(a + h) - g(a)$$
$$= hkD_2D_1f(x_1, y_1).$$

We prove similarly that this is equal to

$$hkD_1D_2f(x_2, y_2)$$

for some x_2 in $\mathbb{M}$ and y_2 in $\mathbb{N}$. Therefore

$$D_2D_1f(x_1, y_1) = D_1D_2f(x_2, y_2).$$

But

$$|D_2D_1f(x_1, y_1) - D_2D_1f(a, b)| < \tfrac{1}{2}c$$

and

$$|D_1 D_2 f(x_2, y_2) - D_1 D_2 f(a, b)| < \tfrac{1}{2}c,$$

and so

$$|D_1 D_2 f(a, b) - D_2 D_1 f(a, b)| < c,$$

which is a contradiction. $\square$

In general, if f is in $\mathbb{R}^n$ into $\mathbb{R}$ and $D_i D_j f$ and $D_j D_i f$ are continuous at α, then

$$D_i D_j f(\alpha) = D_j D_i f(\alpha).$$

The proof is as above.

If we apply the theorem to $D_k f$ (where the third-order derivatives of f are continuous), we have

$$D_i D_j D_k f(\alpha) = D_j D_i D_k f(\alpha).$$

If the third-order derivatives are continuous on a neighbourhood of α, then

$$D_i D_j D_k f(\alpha) = D_i D_k D_j f(\alpha)$$

and so on. Thus we can permute the D_is without affecting the value of the derivative.

Exercises 17. The partial derivatives of the function g in $\mathbb{R}^2$ into $\mathbb{R}$ are continuously differentiable everywhere. For every (x, y) of $\mathbb{R}^2$, $f(x, y) = g(2x + 3y, xy)$. Show that, for every (x, y) of $\mathbb{R}^2$,

$$D_1 D_2 f(x, y) = 6D_1^2 g(2x + 3y, xy) + 3y D_2 D_1 g(2x + 3y, xy)$$
$$+ 2x D_1 D_2 g(2x + 3y, xy) + xy D_2^2 g(2x + 3y, xy)$$
$$+ D_2 g(2x + 3y, xy)$$

and find similar formulae for $D_1^2 f$, $D_2^2 f$ and $D_2 D_1 f$.

18. What do the relations $u = x$, $v = y/x$ change the equation

$$x^2 D_1^2 f(x, y) + 2xy D_1 D_2 f(x, y) + y^2 D_2^2 f(x, y) = 0 \qquad (7)$$

into?

19. Find as general a function f as you can whose second-order derivatives are continuous and for which (7) holds.

Problem 4. Show that if $f(0, 0) = 0$ and $f(x, y) = (x^3 y - xy^3)/(x^2 + y^2)$ whenever $(x, y) \neq (0, 0)$, then $D_1 f(0, 0)$ and $D_2 f(0, 0)$ exist but are unequal.

The ∂ notation

We now introduce another notation—historically, the fir
notation for partial derivatives.

Let f be a function in $\mathbb{R}^2$ into $\mathbb{R}$, and let u and v b
functions into $\mathbb{R}$, so that (u, v) is into $\mathbb{R}^2$. If

$$q = f(u, v) \qquad (8$$

then, with respect to (8), we define

$$\partial q / \partial u \quad \text{to be} \quad D_1 f(u, v)$$

and

$$\partial q / \partial v \quad \text{to be} \quad D_2 f(u, v).$$

Example   2. With respect to the relation
$$q = 2u^2 - 3uv$$
we have
$$\frac{\partial q}{\partial u} = 4u - 3v$$

and

$$\frac{\partial q}{\partial v} = -3u.$$

(Here f is, of course, the function for whic▌
$f(x, y) = 2x^2 - 3xy$ for every x and y, so that the give▌
relation is $q = f(u, v)$.)

It is clear how the definition will extend to functions in $\mathbb{R}$
into $\mathbb{R}$.

Definition   If f is in $\mathbb{R}^n$ into $\mathbb{R}$, if each of $u_1, \ldots, u_n$ is into $\mathbb{R}$, and if

$$q = f(u_1, \ldots, u_n),$$

then with respect to this relation

$$\partial q / \partial u_i \quad \text{denotes} \quad D_i f(u_1, \ldots, u_n). \qquad \quad \text{[}$$

Exercises   20. What are $\partial q / \partial u$ and $\partial q / \partial v$ with respect to
$$q = u \cos v + v^2 \sin u.$$

21. What are $\partial r / \partial x$, $\partial r / \partial y$, $\partial \theta / \partial x$ and $\partial \theta / \partial y$ with respect t▌
$r = (x^2 + y^2)^{1/2}$ and $\theta = \arctan(y/x)$?

22. Let $\mathbb{D}$ be the set of points (a, b) of $\mathbb{R}^2$ for which a is strictl▌
positive, let (x, y) be the identity function on $\mathbb{D}$, and le▌

$r = (x^2 + y^2)^{1/2}$ and $\theta = \arctan(y/x)$. There will be equations expressing x and y as functions of r and θ. Find $\partial x/\partial r$, $\partial y/\partial r$, $\partial x/\partial \theta$, and $\partial y/\partial \theta$ with respect to these equations.

23. Let (x, y) be the identity function on $\mathbb{R}^2$, $q = x^2 - 2xy$, $u = x + y$, and $v = 3x + 4y$.

   (a) Find $\partial q/\partial x$ with respect to $q = x^2 - 2xy$.

   (b) Express q as a function of x and u (with no y involved) and find $\partial q/\partial x$ with respect to this relation.

   (c) Express q as a function of x and v, and find $\partial q/\partial x$ with respect to this relation.

Exercise 23 raises an interesting point. It involves several derivatives $\partial q/\partial x$ with respect to several relations. It would be useful to have a notation for distinguishing between them.

In this particular problem we found that

$$q = 3x^2 - 2xu \tag{9}$$

which is of the form

$$q = g(x, u).$$

This function g is unique: if

$$g(x, u) = h(x, u),$$

then $g = h$. (Proof. Let (a, b) be any element of $\mathbb{R}^2$. Now

$$g(x, u)(a, b - a) = g(x(a, b - a), u(a, b - a))$$
$$= g(a, a + b - a)$$
$$= g(a, b),$$

and similarly for h. Therefore $g(a, b) = h(a, b)$ for every (a, b).)

In this case, then, x and u uniquely determine the function g and therefore uniquely determine $D_1 g(x, u)$, i.e. $\partial q/\partial x$ with respect to (9). We denote this derivative by $(\partial q/\partial x)_u$. Similarly, $(\partial q/\partial x)_v$ denotes $\partial q/\partial x$ with respect to $q = h(x, v)$, where h is the uniquely determined function for which q equals $h(x, v)$. In fact

$$q = \tfrac{5}{2}x^2 - \tfrac{1}{2}vx$$

and so

$$\left(\frac{\partial q}{\partial x}\right)_v = 5x - \tfrac{1}{2}v.$$

Thus

$$\left(\frac{\partial q}{\partial x}\right)_y = 2x - 2y \qquad \left(\frac{\partial q}{\partial x}\right)_u = 6x - 3u$$

$$\left(\frac{\partial q}{\partial x}\right)_v = 5x - \frac{1}{2}v.$$

This is a fairly common situation. Very often the functions $q, u, v, w, \ldots$ are such that there is only one f for which

$$q = f(u, v, w, \ldots) \qquad (10)$$

and then the functions $q, u, v, w, \ldots$ uniquely determine

$$\left(\frac{\partial q}{\partial u}\right)_{v, w, \ldots}.$$

In such situations we can use the subscripts instead of the tag 'with respect to such and such a relation'.

A good example of such a situation occurs in physics. If p, v, and u are the pressure, volume, and temperature of a given specimen of gas, there will be a relation between them (such as Boyle's law or van der Waals' law) by means of which any one can be expressed in terms of the other two:

$$p = P(v, u)$$

$$v = V(u, p)$$

$$u = U(p, v).$$

The functions P, V, and U are uniquely determined by this relation. Then $(\partial p/\partial v)_u$ unambiguously denotes $D_1 P(v, u)$ and so on. This is particularly convenient because many of the partial derivatives have physical interpretations. For example, a rate of change of volume with temperature (per unit volume) is a coefficient of expansion. Thus $v^{-1}(\partial v/\partial u)_p$ is the coefficient of expansion at constant pressure.

Sometimes, even with subscripts, we cannot omit the 'with respect to' tag.

Example   3. Let $w = u + v$ where $v = 2u$. What is $(\partial w/\partial u)_v$?

$$w = u + v, \quad \text{whence} \quad (\partial w/\partial u)_v = 1.$$

Also

$$w = 3u, \quad \text{whence} \quad (\partial w/\partial u)_v = 3$$

and

$$w = 3v/2, \quad \text{whence} \quad (\partial w/\partial u)_v = 0 \text{ etc.}$$

Application to polar coordinates

Our definition is not quite strong enough for one important coordinate system: the polar system. We need to modify it if we want to define partial differentiation with respect to r and θ. (Similar modifications will be needed for polar coordinates in three dimensions and certain other curvilinear coordinates.)

If we use the term 'polar coordinate' in the usual elementary sense we cannot define functions r and θ by saying

'let $r(u, v)$ be the polar distance and $\theta(u, v)$ the polar angle of the point (u, v)'

because $[\sqrt{2}, \pi/4]$ and $[\sqrt{2}, 9\pi/4]$ are both polar coordinates of the point $(1, 1)$ and so $\theta(1, 1)$ is ambiguous: it might be $\pi/4$ or $9\pi/4$.

We could, however, define r and θ as follows:

$$\begin{cases} r(0, 0) = \theta(0, 0) = 0. \\ \text{If } (u, v) \neq (0, 0) \text{ then } (r(u, v), \theta(u, v)) \text{ is} \\ \text{the pair } (s, t) \text{ for which} \\ \quad s > 0, 0 \leq t < 2\pi \\ \quad s \cos t = u, s \sin t = v. \end{cases}$$

This has one disadvantage: θ is discontinuous. To be precise, it is discontinuous at every (u, v) for which $u \geq 0$ and $v = 0$. We could define polar coordinates that are continuous along this half-line if we replace the condition $0 \leq t < 2\pi$ by the condition, say, $-\pi < t \leq \pi$, but then θ would be discontinuous along another half-line.

However, if (u, v) is any pair other than $(0, 0)$ we can define polar coordinates that are continuous on some neighbourhood of (u, v). They will be one-to-one on this neighbourhood, and their inverse will be continuous. We can use this fact to define $(\partial q/\partial r)_\theta$ and $(\partial q/\partial \theta)_r$, where q is a function in $\mathbb{R}^2$ into $\mathbb{R}$ as follows.

Let (u, v) be any pair other than $(0, 0)$ interior to dom q. Let (r, θ) be polar coordinates well behaved on some neighbourhood $\mathbb{U}$ of (u, v) as described above. Let (x, y) be the identity function on $\mathbb{U}$. Then

$$x = r \cos \theta \quad \text{and} \quad y = r \sin \theta \quad \text{on } \mathbb{U}.$$

Let us define f by

$$f(u, v) = q(u \cos v, u \sin v)$$

for each u and v for which the right-hand side exists. Then

$$\begin{aligned}
q &= q(x, y) \\
&= q(r \cos \theta, r \sin \theta) \qquad &\text{on } \mathbb{U} \\
&= f(r, \theta) \qquad &\text{on } \mathbb{U}.
\end{aligned}$$

We define

$$(\partial q / \partial r)_\theta(u, v) \quad \text{to be} \quad D_1 f(r(u, v), \theta(u, v))$$

and

$$(\partial q / \partial \theta)_r(u, v) \quad \text{to be} \quad D_2 f(r(u, v), \theta(u, v)).$$

If r^* and θ^* are another pair of polar coordinates, well behaved on a neighbourhood $\mathbb{V}$ of (u, v), then $r = r^*$ on $\mathbb{U} \cap \mathbb{V}$ and $\theta - \theta^*$ is an integral multiple of 2π, and therefore constant. Therefore

$$D_i f(r^*(u, v), \theta^*(u, v)) = D_i f(r(u, v), \theta(u, v)).$$

Thus $(\partial q / \partial r)_\theta(u, v)$ and $(\partial q / \partial \theta)_r(u, v)$ are well defined. Because these are defined for each (u, v), we have defined the functions $(\partial q / \partial r)_\theta$ and $(\partial q / \partial \theta)_r$. Their domains are subsets of $\mathbb{R}^2 \backslash \{(0, 0)\}$. (Just how large their domains are depends on how well behaved q is.)

From the chain rule it follows that

$$D_1 f(r(u, v), \theta(u, v)) = \left\{ (\cos \theta)\left(\frac{\partial q}{\partial x}\right)_y + (\sin \theta)\left(\frac{\partial q}{\partial y}\right)_x \right\}(u, v)$$

for each (u, v), and so

$$\left(\frac{\partial q}{\partial r}\right)_\theta = (\cos \theta)\left(\frac{\partial q}{\partial x}\right)_y + (\sin \theta)\left(\frac{\partial q}{\partial y}\right)_x.$$

Similarly

$$\left(\frac{\partial q}{\partial \theta}\right)_r = -r(\sin \theta)\left(\frac{\partial q}{\partial x}\right)_y + r(\cos \theta)\left(\frac{\partial q}{\partial y}\right)_x. \qquad \square$$

The ∂ notation is particularly suitable for differentiating a function of a function. If $q = f(u, v)$ where (u, v) is in $\mathbb{R}$ into $\mathbb{R}^2$ then, by corollary 1 on p. 50, under the conditions stated there,

$$q' = D_1 f(u, v)u' + D_2 f(u, v)v'.$$

In the ∂ notation, with respect to $q = f(u, v)$, this becomes

$$q' = \frac{\partial q}{\partial u} u' + \frac{\partial q}{\partial v} v'.$$

Similarly, $q = f(u, v, w)$ yields

$$q' = \frac{\partial q}{\partial u} u' + \frac{\partial q}{\partial v} v' + \frac{\partial q}{\partial w} w'.$$

In general (if u is into $\mathbb{R}^n$), with respect to $q = f(u)$,

$$q' = \sum_{i=1}^{n} \frac{\partial q}{\partial u_i} u_i'.$$

The familiar differentiation formulae are easily translated into the ∂ notation. For example, let f, g, u, and v be differentiable and

$$p = f(u, v) \qquad \text{and} \qquad q = g(u, v),$$

and ∂ derivatives be with respect to these relations. If

$$r = p + q,$$

then

$$\frac{\partial r}{\partial u} = D_1(f + g)(u, v)$$

$$= D_1 f(u, v) + D_1 g(u, v)$$

$$= \frac{\partial p}{\partial u} + \frac{\partial q}{\partial u}.$$

Similarly, if

$$s = p \cdot q$$

then

$$\frac{\partial s}{\partial u} = p \cdot \frac{\partial q}{\partial u} + q \cdot \frac{\partial p}{\partial u}$$

and so on.

Now let us suppose that

$$w = g \circ v \qquad \text{and} \qquad v = f \circ u, \qquad (11)$$

where all the functions involved are continuously differentiable, u is into dom f, which is a subset of $\mathbb{R}^m$, and v is into dom g, which is a subset of $\mathbb{R}^n$. Then, with respect to (11),

$$\frac{\partial w_k}{\partial u_j} = \sum_{i=1}^{n} \left(\frac{\partial w_k}{\partial v_i} \right) \left(\frac{\partial v_i}{\partial u_j} \right)$$

for each relevant j and k. The proof is almost mechanical:

$$\frac{\partial w_k}{\partial u_j} = D_j(g \circ f)_k \circ u \qquad \text{because } w = g \circ f \circ u$$

$$= \left(\sum_{i=1}^{n} D_i g_k \circ f \cdot D_j f_i \right) \circ u$$

$$= \sum_{i=1}^{n} D_i g_k \circ f \circ u \cdot D_j f_i \circ u$$

$$= \sum_{i=1}^{n} D_i g_k \circ v \cdot D_j f_i \circ u$$

$$= \sum_{i=1}^{n} \left(\frac{\partial w_k}{\partial v_i} \right) \left(\frac{\partial v_i}{\partial u_j} \right).$$

Example   4. Let u, v, and w be continuously differentiable functions in $\mathbb{R}^2$ into $\mathbb{R}$ with the same domain and

$$w = f(u, v)$$

where f is continuously differentiable. Then, with respect to this relation, if (x, y) is the identity function on $\mathbb{R}^2$,

$$\frac{\partial w}{\partial x} = \frac{\partial w}{\partial u} \frac{\partial u}{\partial x} + \frac{\partial w}{\partial v} \frac{\partial v}{\partial x} \tag{12}$$

and

$$\frac{\partial w}{\partial y} = \frac{\partial w}{\partial u} \frac{\partial u}{\partial y} + \frac{\partial w}{\partial v} \frac{\partial v}{\partial y}. \tag{13}$$

If (u, v) is one-to-one it will have an inverse, say (p, q). Then

$$x = p(u, v) \qquad \text{and} \qquad y = q(u, v)$$

and with respect to the first of these, if p is continuously differentiable,

$$\frac{\partial x}{\partial u} \frac{\partial u}{\partial x} + \frac{\partial x}{\partial v} \frac{\partial v}{\partial x} = 1$$

and

$$\frac{\partial x}{\partial u} \frac{\partial u}{\partial y} + \frac{\partial x}{\partial v} \frac{\partial v}{\partial y} = 0,$$

as we see by replacing w with x in (12) and (13).

Exercises 24. Let a, b, and c be numbers, (x, y) be the identity function on $\mathbb{R}^2$, and

$$u = x \cos a + y \sin a + b$$
$$v = x \sin a - y \cos a + c.$$

Let w be a continuously differentiable function in $\mathbb{R}^2$ into $\mathbb{R}$. Find $(\partial w/\partial u)_v$ and $(\partial w/\partial v)_u$ in terms of $(\partial w/\partial x)_y$ and $(\partial w/\partial y)_x$, and vice versa.

25. Let $\mathbb{P}$ be the set of all (a, b) of $\mathbb{R}^2$ for which both a and b are positive. Let (p, q) be the identity function on $\mathbb{P}$, and

$$u = p^2 - q^2 \quad \text{and} \quad v = 2pq.$$

Find $(\partial u/\partial p)_q$, $(\partial u/\partial q)_p$, $(\partial v/\partial p)_q$, and $(\partial v/\partial q)_p$. Find also $(\partial p/\partial v)_u$, $(\partial p/\partial u)_v$, $(\partial q/\partial v)_u$, and $(\partial q/\partial u)_v$.

26. If w is a function of u and v, where u and v are functions in $\mathbb{R}^2$ into $\mathbb{R}^2$, find $\partial w/\partial u$ with respect to the given relation in terms of $(\partial w/\partial x)_y$, $(\partial u/\partial x)_y$, $(\partial v/\partial x)_y$, $(\partial w/\partial y)_x$, $(\partial u/\partial y)_x$, and $(\partial v/\partial y)_x$, assuming that all the functions involved are continuously differentiable. Here (x, y) is the identity function on $\mathbb{R}^2$.

27. Let u and v be functions in $\mathbb{R}^2$ into $\mathbb{R}$, f and g be functions in $\mathbb{R}^3$ into $\mathbb{R}$, and (x, y) be the identity on $\mathbb{R}^2$. Let

$$v = f(x, y, u) \quad \text{and} \quad w = g(x, u, v). \tag{14}$$

Prove that

$$\left(\frac{\partial w}{\partial x}\right)_y = \frac{\partial w}{\partial x} + \frac{\partial w}{\partial v}\frac{\partial v}{\partial x} + \frac{\partial w}{\partial v}\frac{\partial v}{\partial u}\left(\frac{\partial u}{\partial x}\right)_y + \frac{\partial w}{\partial u}\left(\frac{\partial u}{\partial x}\right)_y$$

where derivatives without subscripts are with respect to (14). Find a similar formula for $(\partial w/\partial y)_x$.

28. Let

$$p = f(u, v, w) \quad \text{and} \quad w = g(u, v). \tag{15}$$

Find a formula for $\partial p/\partial u$ with respect to

$$p = f(u, v, g(u, v))$$

in terms of the various partial derivatives with respect to (15), assuming that all the functions involved are continuously differentiable.

29(a). Let a, b, c, and d be numbers, where a, b, and c are non-zero. Let

$$au + bv + cw = d. \tag{16}$$

Evaluate

$$\frac{\partial u}{\partial v}\frac{\partial v}{\partial w}\frac{\partial w}{\partial u}$$

where the partial derivatives are with respect to the relations obtained by solving (16) for u, v, and w.

29(b). Find some relation other than (16) between functions u, v, and w which can be solved for each of these functions, and evaluate

$$\frac{\partial u}{\partial v} \frac{\partial v}{\partial w} \frac{\partial w}{\partial u}$$

with respect to these solutions.

30. If $(\partial u/\partial v)_w$ and $(\partial v/\partial u)_w$ are both defined, does it follow that their product is constant with value 1?

31. Find $(\partial x/\partial \theta)_r$ and $(\partial x/\partial r)_\theta$ where these polar derivatives are as explained in the text.

32. If polar derivatives are as explained in the text and w is continuously differentiable on $\mathbb{R}^2 \backslash \{(0, 0)\}$ into $\mathbb{R}^2$, prove that

$$\left(\frac{\partial w}{\partial x}\right)_y^2 + \left(\frac{\partial w}{\partial y}\right)_x^2 = \left(\frac{\partial w}{\partial r}\right)_\theta^2 + \frac{1}{r^2}\left(\frac{\partial w}{\partial \theta}\right)_r^2.$$

Higher-order derivatives in ∂ notation

With respect to $q = f \circ u$,

$$\frac{\partial^2 q}{\partial u_i^2} \qquad \text{denotes} \qquad D_i^2 f \circ u$$

and

$$\frac{\partial^2 q}{\partial u_i\, \partial u_j} \qquad \text{denotes} \qquad D_i D_j f \circ u,$$

and so on.

Example   5. With respect to $q = w^5 v^7$,

$$\frac{\partial q}{\partial w} = 5w^4 v^7$$

$$\frac{\partial^2 q}{\partial w^2} = 20w^3 v^7$$

$$\frac{\partial^2 q}{\partial v\, \partial w} = 35w^4 v^6.$$

Note   If the first-order derivatives of f are continuously differentiable, then $D_i D_j f = D_j D_i f$ and so $\partial^2 q/\partial u_i\, \partial u_j = \partial^2 q/\partial u_j\, \partial u_i$.

Exercises   33. Find $\partial^2 q/\partial u^2$, $\partial^2 q/\partial u\, \partial v$, and $\partial^2 q/\partial v^2$ with respect to $q = \arctan(v/u)$.

34. Find $\partial^2 q/\partial x^2$, $\partial^2 q/\partial y^2$, and $\partial^2 q/\partial x\,\partial y$ in terms of the polar first- and second-order derivatives of q.

35. Let u, v, f, and g have continuous second-order derivatives. Prove that, with respect to

$$w = f(u + av) + g(u - av),$$

we have

$$\frac{\partial^2 w}{\partial v^2} = a^2 \frac{\partial^2 w}{\partial u^2} \qquad (a \text{ is a real number}).$$

36. Let f be in $\mathbb{R}^2$ into $\mathbb{R}$, u and v be in $\mathbb{R}$ into dom f, u, v, and f have continuous second-order derivatives, and

$$w = f(u, v).$$

Prove that with respect to this relation

$$w'' = \frac{\partial w}{\partial u} u'' + \frac{\partial w}{\partial v} v'' + 2 \frac{\partial^2 w}{\partial u\,\partial v} u'v' + \frac{\partial^2 w}{\partial v^2} (v')^2 + \frac{\partial^2 w}{\partial u^2} (u')^2.$$

37. Let u be in $\mathbb{R}^2$ into $\mathbb{R}$ and f in $\mathbb{R}^3$ into $\mathbb{R}$, and let both have continuous second-order derivatives. Let (x, y) be the identity function on $\mathbb{R}^2$ and

$$v = f(x, y, u). \tag{17}$$

Prove that

$$\left(\frac{\partial^2 v}{\partial x^2}\right)_y = \frac{\partial^2 v}{\partial x^2} + 2 \frac{\partial^2 v}{\partial x\,\partial u} \left(\frac{\partial u}{\partial x}\right)_y + \frac{\partial^2 v}{\partial u^2} \left(\frac{\partial u}{\partial x}\right)_y^2 + \frac{\partial v}{\partial u} \left(\frac{\partial^2 u}{\partial x^2}\right)_y$$

where the derivatives without subscripts are with respect to (17).

The matrix of derivatives

If f is in $\mathbb{R}^2$ into $\mathbb{R}$, then $Df(\alpha)$ denotes the pair $[D_1 f(\alpha), D_2 f(\alpha)]$ for every α in the domains of both Df_1 and Df_2. If f is in $\mathbb{R}$ into $\mathbb{R}^2$, then $Df(\alpha)$ denotes

$$\begin{bmatrix} D_1 f_1(\alpha) \\ D_1 f_2(\alpha) \end{bmatrix}.$$

(Note. Here $D_1 f_1$ is simply f_1'. We use the D notation because then this special case is included in the general case to follow.)

The reason why we have written $Df(\alpha)$ as a column in the $\mathbb{R}$-into-$\mathbb{R}^2$ case is so that we can use a rectangular array in the general case. For instance, if f is into $\mathbb{R}^3$ into $\mathbb{R}^2$, then

$$Df(\alpha) = \begin{bmatrix} D_1 f_1(\alpha) & D_2 f_1(\alpha) & D_3 f_1(\alpha) \\ D_1 f_2(\alpha) & D_2 f_2(\alpha) & D_3 f_2(\alpha) \end{bmatrix}$$

and, in general, if f is in $\mathbb{R}^n$ into $\mathbb{R}^m$, then $Df(\alpha)$ is the $m \times n$ matrix whose (i, j)th entry is $D_j f_i(\alpha)$. Thus the $[D_1 f(\alpha), D_2 f(\alpha)]$ defined originally is really a 1×2 matrix rather than just a pair. If f is in $\mathbb{R}$ into $\mathbb{R}$, then $Df(\alpha)$ is the 1×1 matrix $[f'(\alpha)]$, i.e. the number $f'(\alpha)$.

$Df(\alpha)$ is called the **matrix of derivatives** of f at α.

We can now write the formula for differentiating a function of a function as follows:

$$\text{if } v = f \circ u, \text{ then}$$
$$Dv = Df \circ u \cdot Du.$$

(The dot in $Df \circ u \cdot Du$ denotes, of course, matrix multiplication.)

In the case $m = n = p = 1$, if Du never takes the value zero we have

$$\text{if } v = f \circ u, \text{ then}$$
$$Dv/Du = f' \circ u.$$

Exercise　38. If κ is a number and if u, v, and w are continuously differentiable functions in $\mathbb{R}$ into $\mathbb{R}$ such that $uv = \kappa w$, show that $u \cdot Dv + v \cdot Du = \kappa \cdot Dw$. What if u and v are in $\mathbb{R}^n$ into $\mathbb{R}^m$?

Matrices of derivatives have an interesting uniqueness property. Let us illustrate it in $\mathbb{R}^2$. If p, q, r, and s are functions whose domain is a subset $\mathbb{U}$ of $\mathbb{R}^2$ and if

$$p \cdot Dx + q \cdot Dy = r \cdot Dx + s \cdot Dy, \tag{18}$$

where (x, y) is the identity function on $\mathbb{R}^2$, then

$$p = r \quad \text{and} \quad q = s. \tag{19}$$

The proof is quite simple: $Dx = (1, 0)$ and $Dy = (0, 1)$, and therefore (18) is equivalent to $(p, q) = (r, s)$, which at once yields (19).

Similar arguments hold in $\mathbb{R}^n$.

Exercises　39. p, q, u, and v are on $\mathbb{R}^2$ onto $\mathbb{R}$, u and v are continuously differentiable,

$$p \cdot Du + q \cdot Dv = 0,$$

and, for every ξ of $\mathbb{R}^2$,

$$\det \mathrm{D}(u, v)(\xi) \neq 0.$$

Prove that $p = q = 0$.

40. Show that if f is in $\mathbb{R}^2$ into $\mathbb{R}$, if (u, v) is in $\mathbb{R}$ into $\mathrm{dom}\, f$, if all these functions are continuously differentiable, and if

$$w = f(u, v),$$

then

$$\mathrm{D}w = \frac{\partial w}{\partial u}\,\mathrm{D}u + \frac{\partial w}{\partial v}\,\mathrm{D}v.$$

Taylor's theorem

This well-known theorem of elementary calculus has a counterpart in higher dimensions. One form of the elementary Taylor's theorem is as follows.

If $F^{(k)}(x)$ exists whenever x is between zero and h (inclusive), then there is a θ between zero and h for which

$$F(h) = \sum_{r=0}^{k-1} \frac{h^r \cdot F^{(r)}(0)}{r!} + \frac{h^k \cdot F^{(k)}(\theta)}{k!} \qquad (20)$$

The first term is a polynomial in h (the Maclaurin polynomial) and the second term can be regarded as a 'remainder term' showing the difference between $F(h)$ and its $(k-1)$th-degree polynomial approximation.

The result can be generalized to functions in $\mathbb{R}^n$. For simplicity, let us start by considering a function f in $\mathbb{R}^2$. We want a formula analogous to (20), i.e. a formula expressing $f(\delta)$ in terms of a polynomial in δ_1 and δ_2 together with a remainder. More generally, let us find such a formula for $f(\alpha + \delta)$.

Let us suppose that the derivatives of f up to the $(k-1)$th order are continuously differentiable. Define a function F in $\mathbb{R}$ into $\mathbb{R}$ by

$$F(t) = f(\alpha + t\delta) \qquad \text{for each } t.$$

Then

$$F'(t) = \delta_1 \cdot \mathrm{D}_1 f(\alpha + t\delta) + \delta_2 \cdot \mathrm{D}_2 f(\alpha + t\delta), \qquad (21)$$

and we can find F'' similarly. In order to express $F^{(r)}$ in terms of f we define a function D_δ as follows:

$$\mathrm{D}_\delta f = \delta_1 \cdot \mathrm{D}_1 f + \delta_2 \cdot \mathrm{D}_2 f$$

for any function f in $\mathbb{R}^2$ into $\mathbb{R}$. (Thus the domain of D_δ is a set of functions.) We denote $D_\delta(D_\delta)$ by $D_\delta{}^2$, and so on. Then (21) becomes

$$F'(t) = D_\delta f(\alpha + t\delta)$$

and, in general,

$$F^{(r)}(t) = D_\delta{}^r f(\alpha + t\delta). \tag{22}$$

Equation (20) with $h = 1$ gives

$$F(1) = \sum_{r=0}^{k-1} \frac{F^{(r)}(0)}{r!} + \frac{F^{(k)}(\theta)}{k!}$$

which, by (22), is

$$f(\alpha + \delta) = \sum_{r=0}^{k-1} \frac{D_\delta{}^r f(\alpha)}{r!} + \frac{D_\delta{}^k f(\alpha + \theta\delta)}{k!}. \tag{23}$$

This is the required formula, as the first term on the right-hand side is a polynomial in δ_1 and δ_2.

The generalization to functions in $\mathbb{R}^n$ is obvious.

Definition If δ belongs to $\mathbb{R}^n$, then D_δ is the function defined as follows:

$$D_\delta(f) = \sum_{r=0}^{n} \delta_r D_r f$$

for every function f in $\mathbb{R}^n$ into $\mathbb{R}$. □

Taylor's theorem

Let f be a function in $\mathbb{R}^n$ into $\mathbb{R}$, let δ belong to $\mathbb{R}^n$, and let the partial derivatives of f up to the kth order be continuous at $\alpha + \theta\delta$ for every θ in $[0; 1]$. Then there is a number θ in $(0; 1)$ for which (23) holds. □

Note If x is the identity function on $\mathbb{R}^n$, then

$$\frac{\partial f}{\partial x_r} = D_r f.$$

Then

$$D_\delta f = \sum_{r=0}^{n} \delta_r D_r f = \sum_{r=0}^{n} \delta_r \cdot \frac{\partial f}{\partial x_r}.$$

Therefore a common notation for D_δ is

$$\delta_1 \cdot \frac{\partial}{\partial x_1} + \delta_2 \cdot \frac{\partial}{\partial x_2} + \ldots + \delta_n \cdot \frac{\partial}{\partial x_n}.$$

Exercises   41. Write out formula (23) in the cases $k = 1$ and $k = 2$, where f is a function in $\mathbb{R}^2$ into $\mathbb{R}$, without using the symbol D_δ.

42. Write out in full $D_\delta{}^3 g$ where g is a function in $\mathbb{R}^2$ into $\mathbb{R}$ whose third-order derivatives are continuous on dom g and $\alpha = (0, 0)$.

43. If f is on $\mathbb{R}^2$ into $\mathbb{R}$, if its second-order partial derivatives are continuous on a neighbourhood $\mathbb{U}$ of $(0, 0)$, if $f(0, 0) = D_1 f(0, 0) = D_2 f(0, 0) = 0$, if the values of the second-order partial derivatives are smaller than κ everywhere on $\mathbb{U}$, and if δ lies in $\mathbb{U}$, what can be said about the size of $f(\delta)$?

44. Let f be a continuously differentiable function in $\mathbb{R}^2$ into $\mathbb{R}$ with the property that, for some number θ in the interval $(0, \frac{1}{2}\pi)$, $D_1 f(x \cos \theta, x \sin \theta)$ and $D_2 f(x \cos \theta, x \sin \theta)$ are positive for every x between 0 and 1. Prove that

$$f(\cos \theta, \sin \theta) > f(0, 0).$$

Implicit functions

An equation between x and y, say

$$x - y^3 = 0, \tag{24}$$

may be solvable for y. This particular equation is, and the solution is

$$y = x^{1/3}. \tag{25}$$

Another equation that can be solved for y is

$$x + y + y^5 = 0. \tag{26}$$

To solve this equation we define a function q as follows. First, we notice that, for any given x, there is one and only one y such that $y^5 + y = -x$. We then define $q(x)$ to be the number which, when added to its fifth power, yields $-x$. Then

$$y = q(x) \tag{27}$$

is the solution of (26).

We expressed the solution of (24) in terms of a known function (for (25) is the same as $y = p(x)$ where p is the cube-root function), whereas for equation (26) we had to define a new function q. This is not an important difference. After all, how is p defined? It is defined in just the same way as q, but using (24) in place of (26).

We can display these solutions graphically: the graph of the equation $y = x^{1/3}$ is the same as the graph of the equation $x - y^3 = 0$, and the graph of $y = q(x)$ is the same as that of $x + y + y^5 = 0$.

Note   The graph of the equation

$$z = f(x, y)$$

is the set of points (x, y, z) of $\mathbb{R}^3$ for which

$$z = f(x, y).$$

More generally, if h is in $\mathbb{R}^3$ into $\mathbb{R}$, the graph of

$$h(x, y, z) = 0$$

is

$$\{\omega \in \operatorname{dom} h : h(\omega) = 0\}.$$

Probably the best name for this set is the **null set** of h.

In general, the following question arises: given a function f in $\mathbb{R}^2$ into $\mathbb{R}$, can we solve the equation

$$f(x, y) = 0$$

in the form

$$y = q(x)?$$

Of course, this means: can we find a function q in $\mathbb{R}$ into $\mathbb{R}$ such that

$$f(x, y) = 0 \quad \text{if and only if} \quad y = q(x)?$$

In terms of graphs, the question is: can we find a function q such that the graph of $y = q(x)$ is the same as that of $f(x, y) = 0$?

The answer is simple. We can do so if and only if no two points in the same vertical line lie on the graph of $f(x, y) = 0$. In other words, we can do so if and only if f has the following property:

$$\text{if } f(x, y) = f(x, y') = 0,$$
$$\text{then } y = y'.$$

If f has this property, then we define q as follows: for each x for which there exists a y such that $f(x, y) = 0$ there will be only one such y, and we define $q(x)$ to be this y.

We say then that the equation $f(x, y) = 0$ 'defines y implicitly as a function of x'. This phrase is, of course, a figure of speech. It is not y that is defined, but q. However, the phrase is both common and vivid, and now that we have explained what it means we shall not hesitate to use it.

Exercises 45. Which of the following equations define y implicitly as a function of x?

(a) $x - \sin y = 0$;
(b) $x - \arctan y = 0$;
(c) $x - |y| = 0$;
(d) $x - y\,|y| = 0$;
(e) $x^2 + y^2 = 1$;
(f) $x^3 + y^3 = 3xy$.

46. What can you say about the graph of $f(x, y) = 0$ if no two points in the same *horizontal* line lie on it? What can you say about the graph of $f(x, y, z) = 0$ in $\mathbb{R}^3$ if no two points in the same vertical line lie on it?

Most graphs of equations, even quite simple equations, do not have the property quoted above, namely that no two points in the same vertical line lie on the graph. However, most reasonably simple graphs have the property *locally* except at a few exceptional points. For instance, in Fig. 4.1, P is a point on the graph of

$$x^3 + y^3 = 3xy.$$

There is a neighbourhood of P, for example the one shaded

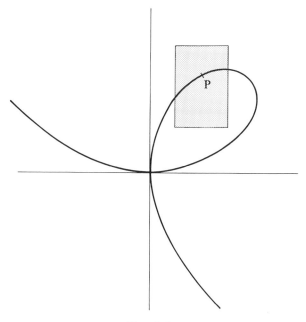

Fig. 4.1

in Fig. 4.1, such that the part of the graph inside the neighbourhood has the property. Thus there is a function q such that the graph of

$$y = q(x)$$

is the part of the graph of $x^3 + y^3 = 3xy$ inside the neighbourhood. If P were *any* point on the graph other than the origin or the point where the tangent is vertical, the same would be true. It is natural to say that $x^3 + y^3 = 3xy$ defines y *locally* as a function of x. Then we make the following definition.

Definition If f is a function in $\mathbb{R}^2$ into $\mathbb{R}$, if $f(a, b) = 0$, if $\mathbb{N}$ is a neighbourhood of (a, b), and if q is a function in $\mathbb{R}$ into $\mathbb{R}$ such that

$$y = q(x)$$

if and only if

$$f(x, y) = 0 \quad \text{and} \quad (x, y) \in \mathbb{N},$$

then we say informally that $y = q(x)$ locally near (a, b). When we say that the equation $f(x, y) = 0$ defines y *locally* as a function of x near (a, b), we mean that such a q exists. □

Example 6. Let (a, b) lie on the graph of $x^2 + y^2 = 1$. Then the following hold:
if $b > 0$, $y = (1 - x^2)^{1/2}$ locally near (a, b);
if $b < 0$, $y = -(1 - x^2)^{1/2}$ locally near (a, b);
if $b = 0$, the equation does not define y locally near (a, b) as a function of x.

Exercise 47. Do the given equations define y locally as a function of x near the given points?
(a) $xy = 1$ near $(1, 1)$;
(b) $\sin y = x$ near $(-1/\sqrt{2}, 5\pi/4)$;
(c) $|y| = x$ near $(1, 1)$, near $(2, 2)$, near $(1, -1)$, or near $(0, 0)$;
(d) $x^3 + y^3 = 3xy$ near $(\frac{3}{2}, \frac{3}{2})$;
(e) $y^3 = x$ near $(0, 0)$ or near $(1, 1)$.

The exercises and examples probably suggest that near a point on the graph of $f(x, y) = 0$ where there is a non-vertical tangent y is locally a function of x. This is not quite true; it may be false if the slope of the tangent varies discontinuously as the point of contact moves, but if we eliminate this possibility (by insisting that f must be continuously differentiable) then the suggested result does hold. This result is called the implicit-function theorem, and we prove it in two steps.

Implicit-function theorem 1

If f is a function in $\mathbb{R}^2$ into $\mathbb{R}$, if $f(a, b) = 0$, if $\mathsf{M} \times \mathsf{N}$ is a neighbourhood of (a, b) contained in dom$f$, if $f(\cdot, y)$ is continuous on M for each y of N, if $f(x, \cdot)$ is continuous on N for each x of M, and if $f(x, \cdot)$ is strictly monotonic on N for each x of M, then there is a neighbourhood U of (a, b) and a function q in $\mathbb{R}$ into $\mathbb{R}$ such that

$$y = q(x) \text{ if and only if } f(x, y) = 0 \text{ and } (x, y) \in \mathsf{U}.$$

Let us first look at the situation diagrammatically (Fig. 4.2). We take the case where $f(a, \cdot)$ is increasing; the case where it is decreasing is treated similarly. The horizontal and vertical sides of the large rectangle in Fig. 4.2 represent M and N. O is the point (a, b). Draw a vertical line through O and on it take a point P above O and a point Q below O. Because the value of f at O is zero and $f(a, \cdot)$ is strictly increasing, f will have a positive value at P and a negative value at Q. Draw a horizontal line through P. By continuity, there will be a segment of this line about P on which f takes positive values; similarly, there will be a horizontal segment through Q on which f takes negative values. Then we can

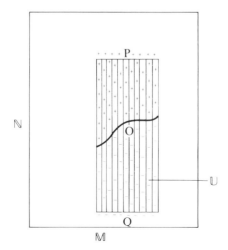

Fig. 4.2

construct the rectangle $\mathbb{U}$ on whose top edge f has positive values and on whose bottom edge f has negative values. Take any vertical line in $\mathbb{U}$. At the bottom the value of f is negative, and it increases continuously up the line until at the top it is positive. Therefore there is just one point on the line where it is zero. The set of all these zero points is the graph of q. Now all we have to do is to formalize this argument.

Proof   Let $c < b < d$ and both c and d belong to $\mathbb{N}$. Then, because $f(a, b) = 0$ and $f(a, \cdot)$ is strictly increasing on $\mathbb{N}$,

$$f(a, c) < 0 < f(a, d).$$

Because $f(\cdot, c)$ is continuous on $\mathbb{M}$ there is a neighbourhood $\mathbb{D}$ of a contained in $\mathbb{M}$ such that $f(x, c) < 0$ whenever x belongs to $\mathbb{D}$. Similarly there is a neighbourhood $\mathbb{E}$ of a contained in $\mathbb{M}$ such that $f(x, d) > 0$ whenever x belongs to $\mathbb{E}$. Then

$$f(x, c) < 0 < f(x, d)$$

whenever x belongs to $\mathbb{D} \cap \mathbb{E}$. For each such $x, f(x, \cdot)$ is continuous, and therefore there is a number x^* between c and d for which

$$f(x, x^*) = 0. \tag{28}$$

Because $f(x, \cdot)$ is strictly monotonic there is only one such number. Then we can define a function q with domain $\mathbb{D} \cap \mathbb{E}$ by letting $q(x)$ be this x^* for each x. Let $\mathbb{U} = (\mathbb{D} \cap \mathbb{E}) \times (c; d)$; this is a neighbourhood of (a, b). If $y = q(x)$, we have $x \in \mathbb{D} \cap \mathbb{E}$ and $c < y < d$, so that (x, y) belongs to $\mathbb{U}$, and, by (28), $f(x, y) = 0$. Conversely, if (x, y) belongs to $\mathbb{U}$ and $f(x, y) = 0$, then x belongs to $\mathbb{D} \cap \mathbb{E}$ and so y must be $q(x)$.   □

Corollary   The q just defined is continuous.

Proof   Let a' belong to $\mathbb{D} \cap \mathbb{E}$ and $b' = q(a')$; let $\mathbb{P}$ be any neighbourhood of b'. Repeat the proof of the theorem but using (a', b') in place of (a, b) and taking c and d in $\mathbb{P} \cap \mathbb{N}$. Whenever x belongs to $\mathbb{D} \cap \mathbb{E}$, $q(x)$ belongs to $(c; d)$ and therefore to $\mathbb{P}$.   □

Exercises   48. If $\mathbb{V}$ is another neighbourhood of (a, b) and r is a function in $\mathbb{R}$ into $\mathbb{R}$ such that $y = r(x)$ if and only if $f(x, y) = 0$ and (x, y) belongs to $\mathbb{V}$, what is the relation between q and r?

49. If $\mathbb{W}$ is a neighbourhood of another point (a', b') for which $f(a', b') = 0$ and s is a function in $\mathbb{R}$ into $\mathbb{R}$ for which $y = s(x)$ if and only if $f(x, y) = 0$ and (x, y) belongs to $\mathbb{W}$, what is the relation between q and s?

Implicit-function theorem 2

If f is a function in $\mathbb{R}^2$ into $\mathbb{R}$, if $f(a, b) = 0$, if f is continuously differentiable on some neighbourhod $\mathbb{L}$ of (a, b) and if $D_2 f(a, b) \neq 0$, then there is a function q in $\mathbb{R}$ into $\mathbb{R}$ and a neighbourhood $\mathbb{U}$ of (a, b) such that

$y = q(x)$ if and only if $f(x, y) = 0$ and $(x, y) \in \mathbb{U}$;

moreover, q is continuously differentiable on a neighbour-hood of a.

Proof   Because $D_2 f(a, b) \neq 0$ and $D_2 f$ is continuous at (a, b), there is a neighbourhood $\mathbb{K}$ of (a, b) in which the values of $D_2 f$ are all of the same sign. Put $\mathbb{M} \times \mathbb{N} = \mathbb{L} \cap \mathbb{K}$; then the conditions of the previous theorem are satisfied, and so the function q exists and, by the corollary, is continuous. For every x near enough to a and every δ near enough to zero, $(x + \delta, q(x + \delta))$ belongs to $\mathbb{U}$. Put $y = q(x)$ and $\epsilon = q(x + \delta) - y$. Then

$$f(x + \delta, y + \epsilon) = f(x + \delta, q(x + \delta)) = 0.$$

Then,

$$0 = f(x + \delta, y + \epsilon) - f(x, y)$$

$$= f(x + \delta, y + \epsilon) - f(x + \delta, y) + f(x + \delta, y) - f(x, y)$$

$$= \epsilon \cdot D_2 f(x + \delta, z) + f(x + \delta, y) - f(x, y)$$

for some z between y and $y + \epsilon$, by the mean-value theorem. $D_2 f(x, y) \neq 0$ and $D_2 f$ is continuous on $\mathbb{L}$; therefore $D_2 f(x + \delta, z)$ is non-zero for every δ which is sufficiently small. Then, for every non-zero δ which is sufficiently small,

$$\frac{q(x + \delta) - q(x)}{\delta} = \frac{\epsilon}{\delta} = -\frac{f(x + \delta, y) - f(x, y)}{\delta \cdot D_2 f(x + \delta, z)}.$$

Because $D_1 f(x, y)$ exists and $D_2 f$ is continuous at (x, y), the limit of the right-hand side as δ approaches zero exists;

therefore $q'(x)$ exists. In fact

$$q'(x) = \frac{-D_1 f(x, y)}{D_2 f(x, y)}$$

$$= \frac{-D_1 f(x, q(x))}{D_2 f(x, q(x))}.$$

The right-hand side is a continuous function of x, and so q' is continuous. □

Note This result means that the tangent at (a, b) to the curve $f(x, y) = 0$ is parallel to

$$D_1 f(a, b) \cdot x + D_2 f(a, b) \cdot y = 0. \tag{29}$$

If, instead of knowing that $D_1 f(a, b) \neq 0$, we know that $D_2 f(a, b) \neq 0$, we can prove the same result (by interchanging the roles of x and y). It follows that if f and g are continuously differentiable on a neighbourhood of (a, b) and if neither $Df(a, b)$ nor $Dg(a, b)$ is $(0, 0)$, then the condition that the curves $f(x, y) = f(a, b)$ and $g(x, y) = g(a, b)$ touch at (a, b) is that (29) is parallel to

$$D_1 g(a, b) \cdot x + D_2 g(a, b) \cdot y = 0.$$

The condition for this is that there is a number λ for which

$$Df(a, b) + \lambda \cdot Dg(a, b) = (0, 0).$$

The theorem extends in the obvious way to n dimensions as follows.

Implicit function theorem for functions in $\mathbb{R}^n$ into $\mathbb{R}$.

If f is a function in $\mathbb{R}^n$ into $\mathbb{R}$, if $f(\alpha) = 0$, if f is continuously differentiable on a neighbourhood of α and if $D_k f(\alpha) \neq 0$, then there is a function q in $\mathbb{R}^{n-1}$ into $\mathbb{R}$ and a neighbourhood $\mathbb{U}$ of α such that

$\xi_k = q(\xi_1, \ldots, \xi_{k-1}, \xi_{k+1}, \ldots, \xi_n)$ if and only if

$f(\xi) = 0$ and $\xi \in \mathbb{U}$.

Moreover, q is continuously differentiable on a neighbourhood of $(\alpha_1, \ldots, \alpha_{k-1}, \alpha_{k+1}, \ldots, \alpha_n)$. □

If $n = 3$, we have an equation analogous to (29) for the tangent-plane at α.

Exercises   50. Show that the equation $x^5 + y^5 + xy = 3$ defines y locally as a function of x near $(1, 1)$.

51. Let $f(x, y)$ be $(x^2 + y^2)^2 - 2x^2 + 2y^2$. Let $f(a, b) = 0$ and $b \neq 0$. Prove that the equation $f(x, y) = 0$ defines y locally near (a, b) as a function of x and, if the function in question is q, find $q'(a)$.

52. Let f be as in Exercise 51. Find the points on the graph of $f(x, y) = 0$ at which the tangent is horizontal and the points at which it is vertical.

53. Find a function f in $\mathbb{R}^2$ into $\mathbb{R}$ and points (a, b) and (a, c) for which $b \neq c$ such that near each of them the equation $f(x, y) = 0$ defines y locally as a function of x.

54. Let f be defined as follows:

$$\begin{cases} f(x, 0) = x \\ f(x, y) = x - \dfrac{y}{2} - y^2 \sin\left(\dfrac{1}{y}\right) \text{ if } y \neq 0. \end{cases}$$

Does the equation $f(x, y) = 0$ define y locally near $(0, 0)$ as a function of x?

55. f is a continuously differentiable function on $\mathbb{R}$ into $\mathbb{R}$. Find a reasonably wide condition that will ensure that the equation

$$3f(xy) - 2f(x) - f(y) = 0$$

defines y locally as a function of x near $(1, 1)$.

Now let us see what happens when we have two equations, say

$$f_1(x, y, z) = 0 \quad \text{and} \quad f_2(x, y, z) = 0.$$

Under what conditions do they define x and y locally as functions of z?

To start, let us suppose that the equations do define x and y locally as functions of z and see what the derivatives of the functions will be. If the solutions are $x = u(z)$ and $y = v(z)$, then

$$f_1(u(z), v(z), z) = 0$$

for every z in the domains of u and v, and so, if $x = u(z)$ and $y = v(z)$,

$$D_1 f_1(x, y, z) \cdot u'(z) + D_2 f_1(x, y, z) \cdot v'(z) + D_3 f_1(x, y, z) = 0$$

and a similar equation holds for f_2. These two equations are linear and so can be solved for $u'(z)$ and $v'(z)$ provided that the determinant of the system is non-zero, i.e.

$$\det\begin{bmatrix} D_1 f_1(x, y, z) & D_2 f_1(x, y, z) \\ D_1 f_2(x, y, z) & D_2 f_2(x, y, z) \end{bmatrix} \neq 0.$$

We might well expect that this is a condition, analogous to the condition $D_2 f(a, b) \neq 0$ in implicit-function theorem 2, under which we shall be able to prove the result we want. This turns out to be true.

Theorem   If f is a function in $\mathbb{R}^3$ into $\mathbb{R}^2$, if $f(a, b, c) = (0, 0)$, if f is continuously differentiable on a neighbourhood of (a, b, c), and if

$$\det\begin{bmatrix} D_1 f_1(a, b, c) & D_2 f_1(a, b, c) \\ D_1 f_2(a, b, c) & D_2 f_2(a, b, c) \end{bmatrix} \neq 0 \qquad (30)$$

then there is a neighbourhood $\mathbb{U}$ of (a, b, c) and a function q in $\mathbb{R}$ into $\mathbb{R}^2$ such that

$$(x, y) = q(z) \text{ if and only if}$$
$$f(x, y, z) = (0, 0) \text{ and } (x, y, z) \in \mathbb{U}.$$

Moreover, q is continuously differentiable.

Proof   By (30), $D_2 f_1(a, b, c)$ and $D_2 f_2(a, b, c)$ are not both zero. Let us suppose that $D_2 f_1(a, b, c) \neq 0$; the proof is similar in the other case. By the implicit-function theorem for functions in $\mathbb{R}^3$ into $\mathbb{R}$, there is a function u and a neighbourhood $\mathbb{L}$ of (a, b, c) such that

$$y = u(x, z) \text{ if and only if}$$
$$f_1(x, y, z) = 0 \text{ and } (x, y, z) \in \mathbb{L}.$$

Moreover

$$D_1 u(a, c) = -D_1 f_1(a, b, c)/D_2 f_1(a, b, c). \qquad (31)$$

Let

$$v(x, z) = f_2(x, u(x, z), z)$$

for every (x, z) in the domain of u. Then

$$D_1 v(a, c) = D_1 f_2(a, b, c) + D_2 f_2(a, b, c) \cdot D_1 u(a, c)$$

and so, by (30) and (31), $D_1 v(a, c) \neq 0$. Therefore, by the implicit-function theorem, there is a neighbourhood $\mathbb{N}$ of

(a, c) and a function q_1 such that

$x = q_1(z)$ if and only if $v(x, z) = 0$ and $(x, z) \in \mathbb{N}$.

Moreover q_1 is continuously differentiable. Let

$$q_2(z) = u(q_1(z), z)$$

for every z in the domain of q_1. Then q_2 is also continuously differentiable. Let

$$\mathbb{U} = \mathbb{L} \cap \{(x, y, z) : (x, z) \in \mathbb{N}\}. \tag{32}$$

Then

$$(x, y) = q(z)$$

if and only if

$$v(x, z) = 0, \ (x, z) \in \mathbb{N}, \text{ and } y = u(x, z),$$

which in turn occurs if and only if

$$f_2(x, y, z) = 0, (x, z) \in \mathbb{N}, f_1(x, y, z) = 0 \text{ and } (x, y, z) \in \mathbb{L}$$

which, by (32), is

$$f(x, y, z) = (0, 0) \qquad \text{and} \qquad (x, y, z) \in \mathbb{U}. \qquad \square$$

It is obvious how the theorem would go for functions in $\mathbb{R}^n$ into $\mathbb{R}^m$, where $m < n$. Instead of quoting the general case, we give an example where $m = 3$ and $n = 5$.

If f is a function in $\mathbb{R}^5$ into $\mathbb{R}^3$, if $f(\alpha) = (0, 0, 0)$, if f is continuously differentiable on a neighbourhood of α, and if $\det[D_1 f(\alpha), D_3 f(\alpha), D_4 f(\alpha)] \neq 0$, then there is a neighbourhood $\mathbb{U}$ of α and a function q in $\mathbb{R}^2$ into $\mathbb{R}^3$ such that

$$(\xi_1, \xi_3, \xi_4) = q(\xi_2, \xi_5)$$

if and only if

$$f(\xi) = (0, 0, 0) \qquad \text{and} \qquad \xi \in \mathbb{U}.$$

To obtain Leibnizian-type formulae from this, we let (x_2, x_5) be the identity on $\mathbb{R}^2$, and (x_1, x_3, x_4) be (q_1, q_2, q_3). We let w denote $f(x_1, x_2, x_3, x_4, x_5)$. Then $w = 0$ on dom q and so, for $i = 1, 2, 3$, with respect to $w = f(x_1, \ldots, x_5)$,

$$\frac{\partial w_i}{\partial x_1}\left(\frac{\partial x_1}{\partial x_2}\right)_{x_5} + \frac{\partial w_i}{\partial x_2} + \frac{\partial w_i}{\partial x_3}\left(\frac{\partial x_3}{\partial x_2}\right)_{x_5} + \frac{\partial w_i}{\partial x_4}\left(\frac{\partial x_4}{\partial x_2}\right)_{x_5} = 0.$$

These three equations can be solved to give the partial derivatives of x_1, x_3, and x_4 with respect to x_2. There are similar equations with x_2 and x_5 interchanged, which yield the partial derivatives with respect to x_5. (Note. Here we have in effect used the second and fifth coordinate variables to parametrize the surface. We would obtain the same formulae if we used any other continuously differentiable parametrization.)

Exercises   56. At what points do the equations

$$x + y + z = 0$$
$$x^3 + y^3 + z^3 = 0$$

define x and y locally as functions of z?

57. Show that in a neighbourhood of $(1, 1, 1)$ the intersection of the graphs of $(x^2 + y^2)z^2 = 2$ and $(x + z)^2 + y^2 = 5$ is a curve with equations of the form

$$x = p(z) \qquad y = q(z).$$

Find the direction ratios of the tangent to the curve at $(1, 1, 1)$.

The inverse-function theorem

Can we solve the equations

$$x = r \cos \theta \qquad y = r \sin \theta$$

for r and θ? A little algebra leads us to

$$r = \pm(x^2 + y^2)^{1/2} \qquad \theta = \arctan\left(\frac{y}{x}\right) + n\pi,$$

an infinity of solutions. However, if $r = 1$ and $\theta = 3\pi/4$, say, then $x = -1/\sqrt{2}$ and $y = 1/\sqrt{2}$, and we might ask for a solution that satisfies the condition

$$r = 1 \text{ and } \theta = 3\pi/4 \text{ when } x = -1/\sqrt{2} \text{ and } y = 1/\sqrt{2}.$$

The answer is

$$r = (x^2 + y^2)^{1/2} \qquad \theta = \arctan\left(\frac{y}{x}\right) + \pi.$$

This does not hold for every x and y; it fails if $x = 0$. It is, however, a perfectly good local solution: there is a neighbourhood of $(-1/\sqrt{2}, 1/\sqrt{2})$ on which it holds.

In analysis we are not interested in the practical details of solving equations like these but in finding general conditions under which solution is possible. Let us investigate the solution for x and y of

$$z = f_1(x, y) \qquad w = f_2(x, y). \tag{33}$$

We suppose that

$$f_1(a, b) = c \quad \text{and} \quad f_2(a, b) = d$$

and ask for a solution

$$x = q_1(z, w) \qquad y = q_2(z, w)$$

of (33) on a neighbourhood $\mathbb{V}$ of (a, b) satisfying

$$q_1(c, d) = a \quad \text{and} \quad q_2(c, d) = b.$$

For q to be a local solution of (33) we need, for each (x, y) of $\mathbb{V}$,

$$x = q_1(z, w) \quad \text{and} \quad y = q_2(z, w)$$

to imply

$$z = f_1(x, y) \quad \text{and} \quad w = f_2(x, y).$$

To sum up, we have a function f in $\mathbb{R}^2$ into $\mathbb{R}^2$ and number pairs (a, b) and (c, d) such that

$$f(a, b) = (c, d),$$

and we want a function q in $\mathbb{R}^2$ into $\mathbb{R}^2$ and a neighbourhood $\mathbb{V}$ of (a, b) such that

$$(x, y) = q(z, w) \text{ and } (x, y) \in \mathbb{V} \text{ imply that } (z, w) = f(x, y).$$

We have here a special case of the implicit-function theorem, as we can see if we define a function u in $\mathbb{R}^4$ into $\mathbb{R}^2$ by setting

$$u(x, y, z, w) = f(x, y) - (z, w) \qquad (34)$$

for every (x, y) in the domain of f and every (z, w) in $\mathbb{R}^2$. Then $u(a, b, c, d) = 0$ and so, by the implicit-function theorem, if u is continuously differentiable on a neighbourhood of (a, b, c, d) and

$$\det \begin{vmatrix} D_1 u_1(a, b, c, d) & D_2 u_1(a, b, c, d) \\ D_1 u_2(a, b, c, d) & D_2 u_2(a, b, c, d) \end{vmatrix} \neq 0$$

there is a neighbourhood $\mathbb{U}$ of (a, b, c, d) and a function q such that

$$(x, y) = q(z, w) \text{ if and only if}$$
$$u(x, y, z, w) = 0 \text{ and } (x, y, z, w) \in \mathbb{U}.$$

Let

$$\mathbb{W} = \{(x, y) : (x, y, f_1(x, y), f_2(x, y)) \in \mathbb{U}\}.$$

Using (34) we see that if f is continuously differentiable on a neighbourhood of (a, b) and

$$\det \mathrm{D}f(a, b) \neq 0,$$

then there is an open set $\mathbb{W}$ containing (a, b) and a function q such that

$$(x, y) = q(z, w) \text{ if and only if}$$
$$f(x, y) = (z, w) \text{ and } (x, y) \in \mathbb{W}.$$

In other words

$$(x, y) = q(w, z) \text{ if and only if } f_{\mathbb{W}}(x, y) = (z, w)$$

and so q is the inverse of $f_{\mathbb{W}}$, a restriction of f. Thus we have proved that a continuously differentiable function is locally invertible wherever its matrix of derivatives is non-singular. (For the neighbourhood $\mathbb{V}$ mentioned above we can take any neighbourhood of (a, b) contained in $\mathbb{W}$.) The same is true for functions in $\mathbb{R}^n$ into $\mathbb{R}^n$ in general. This result is known as the inverse-function theorem and is as follows.

Theorem   If f is a continuously differentiable function in $\mathbb{R}^n$ into $\mathbb{R}^n$ and α belongs to the domain of f and $\det \mathrm{D}f(\alpha) \neq 0$, then there is an open set $\mathbb{V}$ containing α and a function q such that

$$\xi = q(\eta) \text{ if and only if } \eta = f(\xi) \text{ and } \xi \in \mathbb{V}.$$

(Thus q and the restriction of f to $\mathbb{V}$ are mutual inverses.)

Exercises   58. At what points does f have a local inverse if $f(x, y) = (x^2 + y^2, 2xy)$ for every (x, y) in $\mathbb{R}^2$?
59. At what points do the equations $x = r \cos \theta$, $y = r \sin \theta$ define r and θ locally as functions of x and y?

Problem   5. Is the function q of the inverse-function theorem continuous at $f(\alpha)$, differentiable at $f(\alpha)$, continuously differentiable at $f(\alpha)$? If the second-order derivatives of f are continuous on $\mathrm{dom}\, f$, what can you say about q?

Functional dependence

We have just seen that we can solve the equations

$$z = f(x, y) \qquad w = g(x, y)$$

locally for x and y where $\mathrm{D}(f, g)$ is non-singular. Let us turn to the opposite problem: one situation in which we clearly cannot solve these equations is when f is a function of g (or g is a function of f). For instance, trying to solve

$$z = x + y \qquad w = \sin(x + y)$$

for x and y is a hopeless task.

Let us suppose, then, that f is a function of g, say $f = u \circ g$. Now what about the matrix of derivatives? Clearly (if f, g, and u are continuously differentiable)

$$D_1 f = u' \circ g \cdot D_1 g \quad \text{and} \quad D_2 f = u' \circ g \cdot D_2 g,$$

and so $D(f, g)$ is everywhere singular.

We naturally ask what other situations make this matrix everywhere singular, and the answer is that this is essentially the only one. We state this result as a theorem.

Theorem If f and g are continuously differentiable functions in $\mathbb{R}^2$ into $\mathbb{R}$, if

$$\det\begin{bmatrix} D_1 f(x, y) & D_2 f(x, y) \\ D_1 g(x, y) & D_2 g(x, y) \end{bmatrix} = 0 \tag{35}$$

for every (x, y) in some neighbourhood $\mathbb{N}$ of (a, b), and if $D_1 f(a, b) \neq 0$, then there is a function u such that

$$g = u \circ f \quad \text{near} \quad (a, b).$$

Proof Let $c = f(a, b)$. Then, by applying the implicit-function theorem to the equation $z - f(x, y) = 0$, we see that there is a function q and a neighbourhood $\mathbb{U}$ of (a, b, c) such that

$$x = q(y, z) \text{ if and only if}$$

$$z - f(x, y) = 0 \text{ and } (x, y, z) \in \mathbb{U}.$$

Moreover, if $x = q(y, z)$, then

$$D_1 q(y, z) = \frac{-D_2 f(x, y)}{D_1 f(x, y)}. \tag{36}$$

Because q is continuous, the set of all (y, z) in dom q for which $(q(y, z), y)$ lies in $\mathbb{N}$ contains a neighbourhood $\mathbb{M}$ of (b, c). Let $v(y, z) = g(q(y, z), y)$. Then

$$D_1 v(y, z) = D_1 g(q(y, z), y) \cdot D_1 q(y, z) + D_2 g(q(y, z), y)$$

and so, if $(y, z) \in M$ and $x = q(y, z)$ then, by (36),

$$D_1 v(y, z) = D_1 g(x, y) \frac{-D_2 f(x, y)}{D_1 f(x, y)} + D_2 g(x, y)$$

$$= 0 \qquad \text{by (35).}$$

Then

$$v(y, z) = v(b, z) \text{ for every } (y, z) \text{ in } M,$$

i.e.

$$g(x, y) = g(q(b, z), b)$$
$$= u(z) \qquad \text{where } u \text{ is } g(q(b, \cdot), b)$$
$$= u(f(x, y)).$$

This holds for every (x, y) for which $z = f(x, y)$ and $(y, z) \in M$, and because f is continuous this means that it holds for every (x, y) in some neighbourhood of (a, b). $\square$

Exercise 60. In each of the following cases is f a function of g on any neighbourhood? ((x, y) is the identity function on $\mathbb{R}^2$.)
(a) $f = x + y$, $g = x^2 + y^2$.
(b) $f = x$, $g = y$.
(c) $f = (1 - xy)/(x + y)$, $g = (x + y)^2/(1 + x^2)(1 + y^2)$.

Problems 6. (In $\mathbb{R}$.) On what intervals, if any, is cos a function of sin?
7. If f, g, and h are continuously differentiable functions with domain $\mathbb{R}^2$ and values in $\mathbb{R}$, is there necessarily a neighbourhood $\mathbb{N}$ and a function u such that $f(\xi) = u(g(\xi), h(\xi))$ for every ξ in $\mathbb{N}$?

Local maxima and minima

Definitions

A function f in $\mathbb{R}^n$ into $\mathbb{R}$ is said to have a **local maximum** at α if there is a neighbourhood U of α such that

$$f(\xi) \le f(\alpha) \qquad \text{whenever} \qquad \xi \in U.$$

The local maximum is **strict** if $f(\xi) < f(\alpha)$ whenever $\xi \in U$ but $\xi \ne \alpha$. The definition of **local minimum** is similar, with $\ge$ replacing $\le$. An **extremum** is a maximum or a minimum. $\square$

In the case $n = 2$, clearly, if f has a local maximum at α then $f(\cdot, \alpha_2)$ has a local maximum at α_1 and $f(\alpha_1, \cdot)$ has a local maximum at α_2.† This follows immediately from the

† See p. 6 for definitions of $f(\cdot, \alpha_2)$ and $f(\alpha_1, \cdot)$.

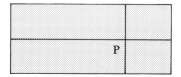

Fig. 4.3

definition, and can be seen from Fig. 4.3; if the value of f at P is the highest in the shaded rectangle, then it is the highest on the horizontal line and also the highest on the vertical line. Because $f(\cdot, \alpha_2)' = D_1 f(\cdot, \alpha_2)$ and $f(\alpha_1, \cdot)' = D_2 f(\alpha_1, \cdot)$, it follows that if f has a local extremum at α and if $Df(\alpha)$ exists, then it must equal $[0, 0]$. Hence we can search for the local extrema of a function in $\mathbb{R}^2$ into $\mathbb{R}$ by computing its derivatives and investigating those points in the domain of f for which $Df(\alpha) = [0, 0]$ or $Df(\alpha)$ fails to exist. Other points can be ruled out.

Similar results hold in the case $n > 2$. If $Df(\alpha) = [0, \ldots, 0]$, then α is a **stationary point** for f. The stationary points, together with those α in the domain of f for which $Df(\alpha)$ fails to exist, are **critical points** for f. Thus only at a critical point can a function have a local extremum.

If f is a twice-differentiable function in $\mathbb{R}$ into $\mathbb{R}$ and α is a stationary point, there is a well-known test for distinguishing between a maximum and a minimum at α, namely the second-order derivative test. There is a similar but more complicated test for functions in $\mathbb{R}^2$ into $\mathbb{R}$. We state it as a theorem.

Theorem Let f be a function from $\mathbb{R}^2$ into $\mathbb{R}$ whose second-order partial derivatives are continuous on some neighbourhood $\mathbb{U}$ of a stationary point α. If

$$D_1^2 f(\alpha) \cdot D_2^2 f(\alpha) - D_1 D_2 f(\alpha)^2 \qquad (37)$$

is positive, then f has a strict local extremum at α, which is a maximum if $D_1^2 f(\alpha)$ is negative and a minimum if it is positive.

Proof Let us suppose that $D_1^2 f(\alpha)$ is negative; the other case is similar. By Taylor's theorem, if $\alpha + \delta \in \mathbb{U}$, then for some θ in $\mathbb{U}$

$$f(\alpha + \delta) = f(\alpha) + \tfrac{1}{2}\delta_1^2 D_1^2 f(\theta)$$
$$+ \delta_1 \delta_2 D_1 D_2 f(\theta) + \tfrac{1}{2}\delta_2^2 D_2^2 f(\theta).$$

By (37), the fact that $D_1^2 f(\alpha) < 0$, and continuity, there is a neighbourhood $\mathbb{W}$ of α such that

$$D_1^2 f(\xi) \cdot D_2^2 f(\xi) - D_1 D_2 f(\xi)^2 > 0 \quad \text{and} \quad D_1^2 f(\xi) < 0$$

whenever $\xi \in \mathbb{W}$. Therefore if $\alpha + \delta \in \mathbb{U} \cap \mathbb{W}$,

$$f(\alpha + \delta) - f(\alpha) = \tfrac{1}{2}(\delta_1^2 a + 2\delta_1 \delta_2 b + \delta_2^2 c)$$

where $ac - b^2 > 0$ and $a < 0$.

Now consider $\delta_1^2 a + 2\delta_1 \delta_2 b + \delta_2^2 c$ as a quadratic in δ_1 and δ_2. It has no real roots and so has the same sign for every δ_1 and δ_2 (except $\delta_1 = \delta_2 = 0$), i.e. the sign of a. Therefore, if $\alpha + \delta \in \mathbb{U} \cap \mathbb{W}$, $f(\alpha + \delta) < f(\alpha)$ unless $\alpha + \delta = \alpha$, and so f has a strict local maximum at α. $\qquad \square$

Corollary If (37) is negative, then f does not have a local extremum at α.

Proof The quadratic now has two real roots and therefore changes sign. Thus there is a δ for which it is positive and, because the sign is not altered if we multiply δ by a non-zero number, there is such a δ in any neighbourhood of $(0, 0)$ and, for such a δ, $f(\alpha + \delta) > f(\alpha)$. Similarly there is, in any neighbourhood of $(0, 0)$, a δ for which $f(\alpha + \delta) < f(\alpha)$. $\qquad \square$

Note To generalize this to the case where f is a function in $\mathbb{R}^n$ into $\mathbb{R}$ we need enough linear algebra to deal with a quadratic in n variables, i.e.

$$\sum \delta_i \delta_j D_i D_j f(\alpha).$$

If this is always positive (except when every δ_i is zero), then f has a local minimum at α. The condition for this is that the matrix $Df(\alpha)$ should be positive definite, and the condition for this is that, for each i from 1 to n, the determinant of the submatrix formed by taking the first i rows and columns should be positive. Of course, the condition for f to have a local maximum is that $-f$ should have a local minimum.

Exercises 61. Find the local maxima and minima of f, where $f(x, y)$ is
(a) $x^2 - xy + y^2$,
(b) $\sin x + \sin y + \sin(x + y)$,
(c) $x^{-1} + xy - 8y^{-1}$,
(d) $x^3 + y^3 - 3xy$,
(e) $xy/(x^2 + y^2)$,
(f) $x^3 y^2(6 - x - y)$.

62. Find a function f and a point α in $\mathbb{R}^2$ such that $Df(\alpha) = 0$, $D_1^2f(\alpha) \cdot D_2^2f(\alpha) - D_1D_2f(\alpha)^2 = 0$, and f has a local maximum at α. Find another function f and a point α in $\mathbb{R}^2$ satisfying the first two conditions but such that f does not have a local extremum at α. What can we say about $D_1^2f(\alpha) \cdot D_2^2f(\alpha) - D_1D_2f(\alpha)^2$ if the second-order derivatives of f are continuous at α and f has a non-strict extremum at α?

63. Find a function f and a point α such that $f(\cdot, \alpha_2)$ has a local maximum at α_1 and $f(\alpha_1, \cdot)$ has a local maximum at α_2, but f does not have a local maximum at α.

64. An engineer wants to find the least value of

$$\frac{8}{x} + \frac{4}{y} + 2xy$$

for all positive x and y. He argues as follows. Let us look at the graph of this function of x and y in the positive quadrant. The expression is (i) large for small x because the term $8/x$ is, (ii) large for small y because the term $4/y$ is, and (iii) large for large x and y because the term $2xy$ is. Thus the graph will have 'mountain ranges' along the axes and points on it that are far distant from the origin will be high, and so the central part will be a 'basin'. Thus it should be possible to find a compact set $\mathbb{S}$ in the positive quadrant such that every point on the graph outside $\mathbb{S}$ is higher than at least one point in $\mathbb{S}$. Because the function is continuous, it will have a minimum on the compact set $\mathbb{S}$. The minimum on $\mathbb{S}$ is the least value of the function.

Is this argument valid? If not, what more is needed to make it valid? If it is (or can be made) valid, complete the solution by specifying a suitable set $\mathbb{S}$ and finding the least value and the point where it occurs.

65. Does $(x + y)^3 - 12xy$ have a least value under the conditions $x > 0$, $y > 0$?

66. Does $(x + y)^3 - 12xy$ have a least value? If so, find it.

67. Let a and b be two given numbers. Why is it obvious that

$$ax + by + (1 - x^2 - y^2)^{1/2}$$

has greatest and least values? Find them.

Problems 8. If f is continuously differentiable everywhere in $\mathbb{R}^n$ and has exactly one stationary point, which is a local maximum, is the value of f there necessarily the greatest value of f?

9. $(0, 0)$ is an interior point of $\operatorname{dom} f$ and every vertical plane section of the graph of $z = f(x, y)$ through the origin has a local minimum there. Does f necessarily have a local minimum there?

Restricted extrema

Let us seek the points on the curve

$$x^2 + xy + y^2 - 1 = 0$$

that are nearest the origin. The square of the distance from the origin to the point (x, y) is

$$x^2 + y^2,$$

and this point will lie on the curve if

$$x^2 + xy + y^2 - 1 = 0.$$

Therefore we are in fact seeking the greatest and least values of the restriction of f, where $f(x, y)$ is

$$x^2 + y^2,$$

to the null set of g, where $g(x, y)$ is

$$x^2 + xy + y^2 - 1.$$

This is a good example of the following problem: find the extrema of the restriction of one continuously differentiable function to the null set of another.

The fact that the extrema of continuously differentiable functions occur only at stationary points is of great help in finding them, and we would naturally like a similar result for our restricted functions.

Let us return to our original example. The graph of

$$x^2 + xy + y^2 - 1 = 0$$

is an ellipse, and the graphs of

$$x^2 + y^2 = \text{constant}$$

are circles (Fig. 4.4). Let us imagine that a point P moves round the ellipse. The value of $x^2 + y^2$ at P changes as P crosses the various circles. Therefore at the point where $x^2 + y^2$ has an extremum P cannot be crossing a circle; it must be moving tangentially to it.

Our general problem is to find the extrema of the restriction of a function f to the null set of g. If the null set is a smooth curve, the arguments above will apply; if our restriction of f has an extremum at α, then the graph of $f(x, y) = f(\alpha)$ cannot cross the graph of $g(x, y) = 0$ at α unless f or g has an extremum at α. Thus (provided that the

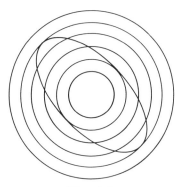

Fig. 4.4

graphs are smooth curves) either

$$Dg(\alpha) = (0, 0)$$

or there is a number λ such that

$$Df(\alpha) + \lambda \cdot Dg(\alpha) = (0, 0).$$

We now give a formal proof of all this.

Theorem Let f and g be continuously differentiable functions in $\mathbb{R}^2$ into $\mathbb{R}$, let α lie in both dom f and dom g, let the restriction of f to the null set of g have an extremum at α, and let $Dg(\alpha) \neq (0, 0)$. Then there is a number λ such that

$$Df(\alpha) + \lambda \cdot Dg(\alpha) = (0, 0).$$

Proof $Dg(\alpha) = (D_1g(\alpha_1, \alpha_2), D_2g(\alpha_1, \alpha_2)) \neq (0, 0)$. Let us suppose that $D_2g(\alpha_1, \alpha_2) \neq 0$. (The proof will be similar if it is $D_1g(\alpha_1, \alpha_2)$ that is not zero.) By the implicit-function theorem, there is a neighbourhood $\mathbb{U}$ of α_1 and a continuously differentiable function q on $\mathbb{U}$ into $\mathbb{R}$ such that

$$q(x) = y \text{ if and only if } g(x, y) = 0 \text{ and } x \in \mathbb{U}.$$

Then $g(x, q(x)) = 0$ whenever $x \in \mathbb{U}$, and so

$$D_1g(x, q(x)) + D_2g(x, q(x)) \cdot q'(x) = 0.$$

In particular,

$$D_1g(\alpha_1, \alpha_2) + D_2g(\alpha_1, \alpha_2) \cdot q'(\alpha_1) = 0. \tag{38}$$

Let p be defined by

$$p(x) = f(x, q(x)).$$

Because $(x, q(x))$ is in the null set of g whenever $x \in U$, p has an extremum at α_1. Therefore $p'(\alpha_1) = 0$, i.e.

$$D_1 f(\alpha_1, \alpha_2) + D_2 f(\alpha_1, \alpha_2) \cdot q'(\alpha_1) = 0. \tag{39}$$

Now put

$$\lambda = -\frac{D_2 f(\alpha)}{D_2 g(\alpha)}. \tag{40}$$

Then

$$
\begin{aligned}
Df(\alpha) + \lambda \cdot Dg(\alpha) &= (D_1 f(\alpha), D_2 f(\alpha)) \\
&\quad + \lambda \cdot (D_1 g(\alpha), D_2 g(\alpha)) \\
&= D_2 f(\alpha) \cdot (-q'(\alpha_1), 1) \\
&\quad + \lambda \cdot D_2 g(\alpha) \cdot (-q'(\alpha_1), 1) \\
&\qquad\qquad\qquad \text{by (38) and (39)} \\
&= (0, 0) \qquad \text{by (40)} \qquad\qquad \square
\end{aligned}
$$

Again, the coordinates of the points on the curve

$$g_1(x, y, z) = g_2(x, y, z) = 0$$

form the null set of (g_1, g_2). In this case the theorem takes the following form.

Theorem Let f be a continuously differentiable function in $\mathbb{R}^3$ into $\mathbb{R}$ and g a continuously differentiable function in $\mathbb{R}^3$ into $\mathbb{R}^2$, let α belong to both $\operatorname{dom} f$ and $\operatorname{dom} g$, let the restriction of f to the null set of g have an extremum at α, and let $Dg(\alpha)$ have rank 2. Then there is a pair (λ_1, λ_2) of numbers such that

$$Df(\alpha) + \lambda \cdot Dg(\alpha) = (0, 0, 0) \qquad\qquad \square$$

The general form of the theorem is as follows.

Theorem Let f be a continuously differentiable function in $\mathbb{R}^n$ into $\mathbb{R}$ and let g be a continuously differentiable function in $\mathbb{R}^n$ into $\mathbb{R}^m$, where $m < n$. Let α belong to both $\operatorname{dom} f$ and $\operatorname{dom} g$, let the restriction of f to the null set of g have an extremum at α, and let $Dg(\alpha)$ have rank m. Then there is a $1 \times m$ matrix λ such that

$$Df(\alpha) + \lambda \cdot Dg(\alpha) = (0, \ldots, 0).$$

Rather than prove the general case, we shall prove the case $n = 5$, $m = 2$, which is completely typical; to write out the proof of any other case, with this one as example, would be sheer routine.

Proof   Because $Dg(\alpha)$ has rank 2, at least one of its 2×2 submatrices is non-singular. Let us suppose that

$$\det \begin{bmatrix} D_4 g_1(\alpha) & D_5 g_1(\alpha) \\ D_4 g_2(\alpha) & D_5 g_2(\alpha) \end{bmatrix} \neq 0 \qquad (41)$$

(The proof is similar if it is one of the other 2×2 submatrices that is non-singular.)

By the implicit-function theorem there is a neighbourhood $\mathbb{U}$ of $(\alpha_1, \alpha_2, \alpha_3)$ and a continuously differentiable function q on $\mathbb{U}$ into $\mathbb{R}^2$ such that

$$q(\xi_1, \xi_2, \xi_3) = (\xi_4, \xi_5)$$

if and only if $g(\xi) = (0, 0)$ and $(\xi_1, \xi_2, \xi_3) \in \mathbb{U}$.

Then if (x, y, z) is the identity function on $\mathbb{R}^3$,

$$g(x, y, z, q_1, q_2) = (0, 0)$$

and so its matrix of derivatives at $(\alpha_1, \alpha_2, \alpha_3)$ is zero:

$$D_j g_i(\alpha) + D_4 g_i(\alpha) \cdot D_j q_1(\alpha_1, \alpha_2, \alpha_3)$$
$$+ D_5 g_i(\alpha) \cdot D_j q_2(\alpha_1, \alpha_2, \alpha_3) = 0 \quad (42)$$

for $j = 1, 2,$ or 3, and $i = 1$ or 2. Let

$$p(\xi_1, \xi_2, \xi_3) = f(\xi_1, \xi_2, \xi_3, q_1(\xi_1, \xi_2, \xi_3), q_2(\xi_1, \xi_2, \xi_3)).$$

Then p has an extremum at $(\alpha_1, \alpha_2, \alpha_3)$ and so $Dp(\alpha_1, \alpha_2, \alpha_3) = 0$, i.e.

$$D_j f(\alpha) + D_4 f(\alpha) \cdot D_j q_1(\alpha_1, \alpha_2, \alpha_3)$$
$$+ D_5 f(\alpha) \cdot D_j q_2(\alpha_1, \alpha_2, \alpha_3) = 0 \quad (43)$$

for $j = 1, 2,$ or 3. By (41) there exist numbers λ_1 and λ_2 such that

$$D_4 f(\alpha) + \lambda_1 \cdot D_4 g_1(\alpha) + \lambda_2 \cdot D_4 g_2(\alpha) = 0$$
$$D_5 f(\alpha) + \lambda_1 \cdot D_5 g_1(\alpha) + \lambda_2 \cdot D_5 g_2(\alpha) = 0. \qquad (44)$$

For each value of i, multiply (42) by λ_i and add these two to (43). Using (44), we see that

$$D_j f(\alpha) + \lambda_1 \cdot D_j g_1(\alpha) + \lambda_2 \cdot D_j g_2(\alpha) = 0$$

for $j = 1, 2,$ or 3. Together with (44) this yields

$$Df(\alpha) + \lambda \cdot Dg(\alpha) = 0. \qquad \square$$

Example 7. Find the greatest and least distance from the origin to the ellipse

$$\begin{cases} x^2 + 2y^2 + 2z^2 = 1 \\ 2x + y + z = 0. \end{cases}$$

We let

$$f(x, y, z) = x^2 + y^2 + z^2$$
$$g_1(x, y, z) = x^2 + 2y^2 + 2z^2 - 1$$
$$g_2(x, y, z) = 2x + y + z,$$

and look for extrema of f on the null set of g. Now

$$Dg(a, b, c) = \begin{bmatrix} 2a & 4b & 4c \\ 2 & 1 & 1 \end{bmatrix},$$

and if this is not of rank 2 we must have

$$a = 4b = 4c$$

which cannot occur on the null set of g. Therefore the conditions of our theorem are satisfied, and so, if (a, b, c) is an extremum,

$$[2a, 2b, 2c] + [\lambda, \mu]\begin{bmatrix} 2a & 4b & 4c \\ 2 & 1 & 1 \end{bmatrix} = (0, 0, 0),$$

i.e.

$$2a + 2a\lambda + 2\mu = 0 \qquad (45)$$

$$2b + 4b\lambda + \mu = 0 \qquad (46)$$

$$2c + 4c\lambda + \mu = 0. \qquad (47)$$

Because (a, b, c) is in the null set of g, we also have

$$a^2 + 2b^2 + 2c^2 = 0 \qquad (48)$$

$$2a + b + c = 0. \qquad (49)$$

If we multiply (45) by a, (46) by b, and (47) by c, and add, we obtain

$$2(a^2 + b^2 + c^2) + 2\lambda = 0$$

and so

$$a^2 + b^2 + c^2 = -\lambda. \qquad (50)$$

The solutions for λ of (45)–(49) are $\lambda = -1/2$ and $\lambda = -3/5$. Then by (50)

$$a^2 + b^2 + c^2 = 1/2 \text{ or } 3/5.$$

These are the only possible extrema. However, because f is continuous and the null set of g is compact, f will have greatest and least values on this set, and so one of our two possible extrema must be the greatest value and one the least. Thus the distances required are $\sqrt{(1/2)}$ and $\sqrt{(3/5)}$.

This technique is due to J. L. Lagrange (1736–1813), and the number we have denoted by λ is called a Lagrange multiplier.

Exercises
68. Find the extrema of $x^2 + y^2$ on the null set of $ax^2 + 2bxy + cy^2 - 1$ where $ac \neq b^2$.

69. Find the greatest and least values of $2x^2 + 2y^2 + z^2$ on the curve

$$x^2 + y^2 + z^2 = 4$$
$$x + y + 2z = 0.$$

70. Find the tangent to the ellipse $5x^2 - 6xy + 5y^2 = 4$ that is furthest from the origin.

71. Does $x + y + z$ have a least value under the condition $1/x + 4/y + 9/z = 1$? If so, find it.

72. Find the volume of the largest rectangular box that has its corners on the ellipsoid $x^2/a^2 + y^2/b^2 + z^2/c^2 = 1$ and its edges parallel to the coordinate axes.

73. Find the greatest and least values, if any, of $a^2x^2 + b^2y^2 + c^2z^2$ on the ellipsoid $x^2/p^2 + y^2/q^2 + z^2/r^2 = 1$.

74. Find the greatest and least values, if any, of $3x^2 + 6xy - 13y^2$ on the curve $-13x^2 + 6xy + 3y^2 = 2$.

75. Find the greatest and least values, if any, of $3x^2 + 6xy + 7y^2$ on the curve $7x^2 + 6xy + 3y^2 = 12$.

76. Find the greatest and least values, if any, of $x^2 + y^2 + z^2$ under the conditions $x^2 + y^2 + z^2 + yz + zx + xy = 1$ and $x + 2y - 2z = 0$.

77. Find the point or points, if any, that are nearest to and furthest from the origin on the curve of intersection of the cone $z^2 = x^2 + y^2$ with each of the following planes: (a) $z = 2x + 3$; (b) $2z = x + 3$.

78. Find the greatest and least values, if any, of $(y - z)(z - x)(x - y)$ under the condition $x^2 + y^2 + z^2 = 1$.

79. Find the greatest and least values, if any, of $x^3 + y^3 + z^3$ under the conditions $x^2 + y^2 + z^2 = 1$ and $x + y + z = 1$.

80. Find the greatest and least values, if any, of $x^3 + y^3 + z^3$ under the conditions $x^2 + y^2 + z^2 = 1$ and $x + y + z = 0$.

Problems
10. Let f and g be continuously differentiable functions in $\mathbb{R}^2$ into $\mathbb{R}$. What happens if we try to find the extrema of the restriction of

f to the null set of g as follows. Solve the equation $g(x, y) = 0$ for y, substitute the solution, say $y = q(x)$, into f, and find the extrema of $f(x, q(x))$ using the simple techniques for functions of one variable.

11. In Problem 10, let $f(x, y)$ be $x^2 + y^2$ and $g(x, y)$ be $x^2 - y$. If we solve $x^2 - y = 0$ for x and substitute in f we obtain

$$f(x, y) = x^2 + y^2 = y + y^2$$

whose least value is $-1/4$. Can the least value of $x^2 + y^2$ under the condition $x^2 = y$ be $-1/4$? If not, what precisely was wrong with our argument?

12. In Problem 10 let $\mathbb{D}$ be the null set of g and let f^* be defined by

$$f^*(\xi) = f(\xi, q(\xi)).$$

Is the greatest value of $f_{\mathbb{D}}$ the greatest value of f^*? Is every stationary value of $f_{\mathbb{D}}$ a stationary value of f^*? Is every local maximum of $f_{\mathbb{D}}$ a local maximum of f^*? (Note. $f_{\mathbb{D}}$ is the restriction of f to $\mathbb{D}$.)

Differentiability

So far we have not said what we mean by 'f is differentiable at α' in the case where f is in $\mathbb{R}^2$ into $\mathbb{R}$. We might be tempted to say that f is differentiable at α if $D_1 f(\alpha)$ and $D_2 f(\alpha)$ exist, but if we made this definition we would find that differentiable functions would lack certain desirable properties. For instance, a function could then be differentiable at α without being continuous there, as we see from the answer to Problem 2 on p. 49. We therefore use a different definition, which in fact is suggested by the geometrical problem of finding a tangent plane to the graph of f.

Let us, then, find the condition that the plane with equation

$$z = lx + my + n$$

should touch the graph of f at the point A with coordinates

$$(\alpha_1, \alpha_2, f(\alpha)).$$

The point P on the plane with x and y coordinates $\alpha + \delta$ has z coordinate

$$l(\alpha_1 + \delta_1) + m(\alpha_2 + \delta_2) + n.$$

However, because the plane passes through A

$$f(\alpha) = l\alpha_1 + m\alpha_2 + n.$$

By eliminating n we find that the z coordinate of P is

$$f(\alpha) + l\delta_1 + m\delta_2.$$

The point Q on the graph vertically in line with P has z coordinate

$$f(\alpha + \delta).$$

Then the distance PQ is

$$f(\alpha + \delta) - f(\alpha) - l\delta_1 - m\delta_2$$

and the angle PAQ approaches zero as P approaches A if and only if

$$\frac{f(\alpha + \delta) - f(\alpha) - l\delta_1 - m\delta_2}{|\delta|}$$

does so. ($|\delta|$ is of course the horizontal distance between A and P.) We therefore make the following definition.

Definition   If f is in $\mathbb{R}^2$ into $\mathbb{R}$, then f is **differentiable** at α if there is a λ in $\mathbb{R}^2$ such that for any positive ϵ there is a positive κ such that

$$\frac{|f(\alpha + \delta) - f(\alpha) - \lambda \cdot \delta|}{|\delta|} < \epsilon$$

whenever $|\delta| < \kappa$.   □

With this definition and a precise definition of tangent plane we could prove that if f is continuous at α, the graph of f has a non-vertical tangent plane at $(\alpha_1, \alpha_2, f(\alpha))$ if and only if f is differentiable at α.

It is clear what λ must be. In the special case $\delta_2 = 0$ we have

$$\frac{|f(\alpha_1 + \delta_1, \alpha_2) - f(\alpha_1, \alpha_2) - \lambda_1\delta_1|}{|\delta_1|} \to 0 \quad \text{as} \quad \delta_1 \to 0$$

and so $\lambda_1 = D_1 f(\alpha)$. A similar argument holds for λ_2, and so $\lambda = Df(\alpha)$.

It is not difficult to prove that if f is differentiable at α then it is continuous at α, and that if f is continuously differentiable at α then it is differentiable at α, but not vice versa. Thus differentiability is a weaker property than continuous differentiability. It is strong enough for some purposes; for example, the formula for differentiating a function of a function can be proved on the assumption that the functions are merely differentiable, not continuously differentiable, at the appropriate points. However, other theorems require continuous differentiability; we shall therefore not go into the question of mere differentiability beyond proving that a continuously differentiable function satisfies the condition for differentiability.

Theorem If f is in $\mathbb{R}^2$ into $\mathbb{R}$ and is continuously differentiable at α, and if $\lambda = (D_1 f(\alpha), D_2 f(\alpha))$, then for any positive ϵ there is a positive κ such that

$$|f(\alpha + \delta) - f(\alpha) - \lambda \cdot \delta| < \epsilon \, |\delta|$$

whenever $|\delta| < \kappa$.

Proof Let $\epsilon > 0$. Because $D_1 f$ and $D_2 f$ are continuous at α there is a κ such that, for $i = 1$ and $i = 2$,

$$|D_i f(\alpha + \delta) - D_i f(\alpha)| < \tfrac{1}{2}\epsilon \qquad \text{whenever} \qquad |\delta| < \kappa.$$

By the mean-value theorem of elementary calculus applied to $f(\cdot, \alpha_2 + \delta_2)$ and to $f(\alpha_1, \cdot)$,

(i) $f(\alpha_1 + \delta_1, \alpha_2 + \delta_2) - f(\alpha_1, \alpha_2 + \delta_2)$
$$= \delta_1 D_1 f(\alpha_1 + \sigma_1, \alpha_2 + \delta_2)$$

and

(ii) $f(\alpha_1, \alpha_2 + \delta_2) - f(\alpha_1, \alpha_2) = \delta_2 D_2 f(\alpha_1, \alpha_2 + \tau_2)$

where $|\sigma_1| \leq |\delta_1|$ and $|\tau_2| \leq |\delta_2|$. For convenience set $\sigma_2 = \delta_2$ and $\tau_1 = 0$. Then $|\sigma| \leq |\delta|$, $|\tau| < |\delta|$, and adding (i) to (ii) yields

$$f(\alpha + \delta) - f(\alpha) = \delta_1 D_1 f(\alpha + \sigma) + \delta_2 D_2 f(\alpha + \tau)$$

Therefore, whenever $|\delta| < \kappa$,

$$|f(\alpha + \delta) - f(\alpha) - \lambda \cdot \delta|$$
$$= |\delta_1 [D_1 f(\alpha + \sigma) - D_1 f(\alpha)] + \delta_2 [D_2 f(\alpha + \tau) - D_2 f(\alpha)]|$$
$$< \tfrac{1}{2}\delta_1 \epsilon + \tfrac{1}{2}\delta_2 \epsilon$$
$$\leq \epsilon \, |\delta|. \qquad \qquad \square$$

Note The theorem can be generalized to the case where f is in $\mathbb{R}^n$ into $\mathbb{R}$ and λ is
$$(D_1 f(\alpha), \ldots, D_n f(\alpha)).$$

If we use the notation for the matrix of derivatives (p. 66), then this is written as

$$Df(\alpha)$$

and regarded as a $1 \times n$ matrix. We can then write the scalar product

$$(D_1 f(\alpha), \ldots, D_n f(\alpha)) \cdot \delta$$

as the matrix product

$$Df(\alpha) \cdot \delta^*$$

where δ^* is the transpose of δ.

We can generalize further: if f is in $\mathbb{R}^n$ into $\mathbb{R}^m$ we can apply the above result to each component of f. We end up with the following theorem (in which $|\xi^*|$ means simply $|\xi|$).

Theorem   If f is in $\mathbb{R}^n$ into $\mathbb{R}^m$ and is continuously differentiable on a neighbourhood of α, and if $\epsilon > 0$, then there is a positive κ such that

$$|f(\alpha + \delta)^* - f(\alpha)^* - Df(\alpha) \cdot \delta^*| < \epsilon \cdot |\delta|$$

whenever $|\delta| < \kappa$.   □

This theorem draws our attention to the function l defined by

$$l(\tau)^* = Df(\alpha) \cdot \tau^* \qquad \text{for every } \tau \text{ of } \mathbb{R}^n.$$

l is a linear function, and is clearly the closest linear approximation to f at α. It is called the 'Fréchet derivative' of f at α.

We can find a relation between the Fréchet derivatives of mutually inverse functions.

Theorem   If f is continuously differentiable and is in $\mathbb{R}^n$ into $\mathbb{R}^n$ and has a continuously differentiable inverse g, then the Fréchet derivatives of f and g are each other's inverses.

Proof   If $\beta = f(\alpha)$ then, by the chain rule,

$$Dg(\beta) \cdot Df(\alpha) = D(g \circ f)(\alpha)$$
$$= Dx(\alpha)$$
$$= I$$

where x is the identity function on $\mathbb{R}^n$ and I is the identity matrix. The Fréchet derivatives l and m are defined by

$$l(\tau)^* = Df(\alpha) \cdot \tau^*$$
$$m(\tau)^* = Dg(\beta) \cdot \tau^*.$$

Then

$$m(l(\tau))^* = Dg(\beta) \cdot l(\tau)^*$$
$$= Dg(\beta) \cdot Df(\alpha) \cdot \tau^*$$
$$= \tau^*.$$

Therefore $m(l(\tau)) = \tau$ and, similarly, $l(m(\tau)) = \tau$. □

While we are on the subject of linear functions, let us prove the following purely algebraic theorem that we shall use later.

Theorem   If $\alpha > 0$ and Λ is a 2×2 matrix whose coefficients all satisfy

$$|\lambda_{ij}| \le \alpha$$

and if

$$l(\tau)^* = \Lambda \cdot \tau^* \qquad \text{for every } \tau \text{ of } \mathbb{R}^2$$

then

$$|l(\tau)| \le 4\alpha \cdot |\tau| \qquad \text{for every } \tau \text{ of } \mathbb{R}^2.$$

Proof $|l(\tau)| \le |\lambda_{11}| \cdot |\tau_1| + |\lambda_{12}| \cdot |\tau_2| + |\lambda_{21}| \cdot |\tau_1| + |\lambda_{22}| \cdot |\tau_2|$
$\le 4\alpha \cdot |\tau|.$

(Note. The theorem clearly holds for $m \times n$ matrices if we replace 4 by mn.) □

We now prove a lemma that we shall need later.

Lemma   Let g be in $\mathbb{R}^2$ into $\mathbb{R}^2$, dom g be open, $\mathbb{K}$ be a compact subset of dom g, and g be continuously differentiable on $\mathbb{K}$. If ϵ is any positive number, there is a positive number δ such that

$$|g(\xi + \eta)^* - g(\xi)^* - Dg(\xi) \cdot \eta^*| < \epsilon \cdot |\eta|$$

whenever $\xi \in \mathbb{K}$, $\xi + \eta \in \mathbb{K}$, and $|\eta| < \delta$.

Proof   Because the $D_j g_i$ are continuous and $\mathbb{K}$ is compact, there is a δ such that

$$|D_j g_i(\xi + \eta) - D_j g_i(\xi)| < \tfrac{1}{4}\epsilon \qquad (51)$$

whenever $\xi \in \mathbb{K}$, $\xi + \eta \in \mathbb{K}$, and $|\eta| < \delta$. Then, by Taylor's

theorem,
$$g_1(\xi + \eta) - g_1(\xi_1, \xi_2 + \eta_2) = \eta_1 \cdot D_1 g_1(\sigma)$$
and
$$g_1(\xi_1, \xi_2 + \eta_2) - g_1(\xi) = \eta_2 \cdot D_2 g_1(\tau)$$

for some σ and τ within $|\eta|$ of ξ. Then

$$|g_1(\xi + \eta) - g_1(\xi) - Dg_1(\xi) \cdot \eta^*|$$
$$= |\eta_1 \cdot \{D_1 g_1(\sigma) - D_1 g_1(\xi)\} + \eta_2 \cdot \{D_2 g_1(\tau) - D_2 g_1(\xi)\}|$$
$$< \tfrac{1}{2}\epsilon \cdot |\eta|.$$

The same applies to g_2, and our result follows. (Note. The lemma can clearly be generalized to functions in $\mathbb{R}^n$ into $\mathbb{R}^m$.) $\qquad\square$

5
Multiple integration

Multiple integration is integration of functions in $\mathbb{R}^n$ into $\mathbb{R}$. It is a fairly straightforward generalization of integration of functions in $\mathbb{R}$ into $\mathbb{R}$. The applications are similar except that they are n dimensional instead of being confined to one dimension: just as the integral of the linear density of a rod gives the mass of the rod, so the integral of the surface density of a plate gives the mass of the plate, and in three dimensions the integral of the density of a body gives the mass of the body. The (Riemann–Darboux) integral of a function in $\mathbb{R}$ into $\mathbb{R}$ is defined as the common bound of certain upper sums and lower sums; so is the multiple integral, the calculation of the sums being n dimensional.

Nearly all the extra difficulties in multiple integration are present in the case $n = 2$. Once they have been overcome, the extension to higher values of n is routine. We therefore present the case $n = 2$ in detail and leave the higher-dimensional case to the reader.

One of the extra difficulties presented by integration in more than one dimension is that we need to be able to integrate over more general sets than just intervals. We therefore define the integral first over intervals, and then extend the theory to cover general bounded sets, and finally to cover certain unbounded sets (and certain unbounded functions).

Partitions

Definitions

The **content** of the compact interval $[a; b]$ in $\mathbb{R}$ is $b - a$. We denote the content of $\mathbb{I}$ by $c(\mathbb{I})$.

The **content** of the compact interval $\mathbb{I} \times \mathbb{J}$ in $\mathbb{R}^2$ is $c(\mathbb{I}) \cdot c(\mathbb{J})$.

If $c(\mathbb{I}) = c(\mathbb{J})$, then $\mathbb{I} \times \mathbb{J}$ is **square**. □

Note If we represent points diagrammatically in the usual way, then a compact interval in $\mathbb{R}$ is represented by a line segment and its content by the length of the segment. A compact interval in $\mathbb{R}^2$ is represented by a rectangle, and its content by the area of the rectangle. Thus we can regard content in $\mathbb{R}$ as a kind of length and content in $\mathbb{R}^2$ as a kind of area (and the obvious analogue in $\mathbb{R}^3$ as a kind of volume).

Definitions Two sets **overlap** if they have an interior point in common. A **partition** of the compact interval $\mathbb{I}$ in $\mathbb{R}$ is a finite set of non-overlapping compact intervals, none of which consists of only a single point, whose union is $\mathbb{I}$. If $\mathcal{P}$ and $\mathcal{Q}$ are partitions, then $\mathcal{P} \times \mathcal{Q}$ denotes

$$\{\mathbb{S} \times \mathbb{T} : \mathbb{S} \in \mathcal{P}, \mathbb{T} \in \mathcal{Q}\}.$$

A **partition** of the compact interval $\mathbb{I} \times \mathbb{J}$ in $\mathbb{R}^2$ is a set of the form $\mathcal{P} \times \mathcal{Q}$ where $\mathcal{P}$ is a partition of $\mathbb{I}$ and $\mathcal{Q}$ is a partition of $\mathbb{J}$.

The intervals that make up a partition are called its **cells**.

The **mesh** of a partition in $\mathbb{R}$ is the greatest of the contents of any of its cells. The **mesh** of $\mathcal{P} \times \mathcal{Q}$ is the larger of the mesh of $\mathcal{P}$ and the mesh of $\mathcal{Q}$. □

Examples 1. Figure 5.1(a) shows a partition $\mathcal{P}$ of $[0; 1]$. Its mesh is $1/5$.
2. Figure 5.1(b) shows a partition $\mathcal{Q}$ of $[-1; 1]$. Its mesh is $1/2$.
3. Figure 5.1(c) shows the partition $\mathcal{P} \times \mathcal{Q}$ of $[0; 1] \times [-1; 1]$. Its mesh is $1/2$.
4. Figure 5.1(d) shows a set of non-overlapping compact intervals that cover $\mathbb{I} \times \mathbb{J}$ but do not form a partition.
5. Clearly any two-dimensional partition can be pictured as a grid, like the one in Fig. 5.1(c).
6. If ξ and η belong to the same cell of a partition in $\mathbb{R}^2$ of mesh k, then $|\xi - \eta| \leq k\sqrt{2}$.

Definition $\mathcal{P}$ is a **refinement** of $\mathcal{Q}$ if each cell of $\mathcal{P}$ is a subset of some cell of $\mathcal{Q}$. To refine a two-dimensional partition diagrammatically, draw extra grid lines.

If $\mathcal{P}$ and $\mathcal{Q}$ are partitions of the same compact interval in $\mathbb{R}^2$, there are partitions that are refinements of both $\mathcal{P}$ and

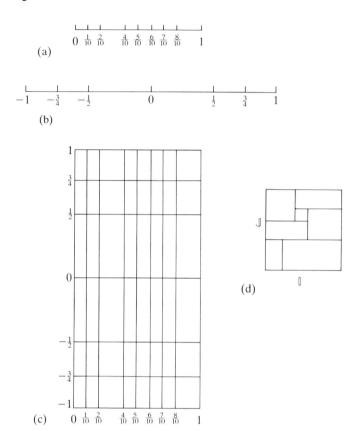

Fig. 5.1

$\mathcal{2}$. To find one such partition, draw the grid lines for $\mathcal{P}$ together with those for $\mathcal{2}$.

If $\mathcal{P}$ is a partition of $\mathbb{I}$, $\sum_{\mathbb{S} \in \mathcal{P}} c(\mathbb{S}) = c(\mathbb{I})$.

(A strict proof for the one-dimensional case, if required, can be devised by induction over the number of cells. The two-dimensional case follows immediately.)

Exercise   1. Let $\mathcal{P}_n$ be the partition of $[0; 1]$ into n cells, each of content $1/n$. Let f be a bounded function with domain $[0; 1]$ and m_r be its supremum on the rth cell. Let

$$\Sigma_n = \sum_{r=1}^{n} m_r / n.$$

(a) Prove that if p is a multiple of n, then $\Sigma_p \le \Sigma_n$.

(b) Is it necessarily true that $\Sigma_3 \le \Sigma_2$?

Integration on compact intervals

Throughout the remainder of this chapter, f will denote a bounded function whose domain is a compact interval in $\mathbb{R}^2$ and whose values are in $\mathbb{R}$. All partitions mentioned will be partitions of $\operatorname{dom} f$.

Definition   Let $\mathscr{P}$ be a partition of $\operatorname{dom} f$. Then

$$\sum_{\mathbb{S} \in \mathscr{P}} c(\mathbb{S}) \sup_{\mathbb{S}} f$$

is the **upper sum** of f for $\mathscr{P}$ and is denoted by $^*\!\sum_{\mathscr{P}} f$, and

$$\sum_{\mathbb{S} \in \mathscr{P}} c(\mathbb{S}) \inf_{\mathbb{S}} f$$

is the **lower sum** of f for $\mathscr{P}$ and is denoted by $_*\!\sum_{\mathscr{P}} f$.   □

Examples   7. If f is constant with value k, $^*\!\sum_{\mathscr{P}} f = {}_*\!\sum_{\mathscr{P}} f = k \cdot c(\operatorname{dom} f)$.

8. If

$$f(x, y) = \begin{cases} 0 \text{ if } x \text{ and } y \text{ are rational} \\ 1 \text{ if not} \end{cases}$$

then $^*\!\sum_{\mathscr{P}} f = c(\operatorname{dom} f)$ and $_*\!\sum_{\mathscr{P}} f = 0$.

9. If $\operatorname{dom} f = [0; 1] \times [0; 1]$ and $f(x, y) = x$ for each x and y, if $\mathscr{P}$ is the partition of $[0; 1]$ consisting of n equal intervals, and if $\mathscr{Q}$ is the partition of $[0; 1]$ consisting of one interval, and if $\mathscr{R} = \mathscr{P} \times \mathscr{Q}$, then

$$^*\!\sum_{\mathscr{R}} f = \sum_{i=1}^{n} \frac{i}{n^2} = \frac{1}{2} + \frac{1}{2n}$$

and

$$_*\!\sum_{\mathscr{R}} f = \sum_{i=1}^{n} \frac{i-1}{n^2} = \frac{1}{2} - \frac{1}{2n}.$$

Lemma   If $\mathscr{Q}$ is a refinement of $\mathscr{P}$, then

$$^*\!\sum_{\mathscr{Q}} f \leq {}^*\!\sum_{\mathscr{P}} f$$

and

$$_*\!\sum_{\mathscr{Q}} f \geq {}_*\!\sum_{\mathscr{P}} f.$$

Proof   Let $\mathbb{S}$ be a cell of $\mathscr{P}$. It will be the union of a number of cells, say $\mathbb{S}_1 \ldots \mathbb{S}_k$, of $\mathscr{Q}$. Because $\mathbb{S}_i \subseteq \mathbb{S}$, we have

$$\sup_{\mathbb{S}_i} f \leq \sup_{\mathbb{S}} f.$$

Then

$$c(\mathbb{S}) \sup_{\mathbb{S}} f = \sum_i c(\mathbb{S}_i) \sup_{\mathbb{S}} f$$

$$\geq \sum_i c(\mathbb{S}_i) \sup_{\mathbb{S}_i} f.$$

If we sum over all the cells $\mathbb{S}$ of $\mathcal{P}$ we obtain

$$^*\!\sum_{\mathcal{P}} f \geq {}^*\!\sum_{\mathcal{Q}} f.$$

The proof for the lower sums is similar. □

Theorem If $\mathcal{P}$ and $\mathcal{R}$ are any partitions of dom f, then

$$_*\!\sum_{\mathcal{R}} f \leq {}^*\!\sum_{\mathcal{P}} f.$$

('Upper sums are greater than lower sums.')

Proof Let $\mathcal{Q}$ be a refinement of both $\mathcal{P}$ and $\mathcal{R}$. Then

$$_*\!\sum_{\mathcal{R}} f \leq {}_*\!\sum_{\mathcal{Q}} f \leq {}^*\!\sum_{\mathcal{Q}} f \leq {}^*\!\sum_{\mathcal{P}} f.$$ □

Corollary

$$\sup {}_*\!\sum_{\mathcal{P}} f \leq \inf {}^*\!\sum_{\mathcal{P}} f$$

where the supremum and infimum are over all partitions $\mathcal{P}$ of dom f. □

Definitions

$\sup {}_*\!\sum_{\mathcal{P}} f$ over all partitions $\mathcal{P}$ of dom f is called the **lower integral** of f and is denoted by $_*\!\int f$, and $\inf {}^*\!\sum_{\mathcal{P}} f$ is the **upper integral** and is denoted by $^*\!\int f$.

If $_*\!\int f = {}^*\!\int f$, then f is said to be **integrable**; the common value of the upper and lower integrals of f is the **integral** of f and is denoted by $\int f$. It follows immediately that f is integrable if and only if, for each positive ϵ, there is a partition $\mathcal{P}$ such that

$$^*\!\sum_{\mathcal{P}} f - {}_*\!\sum_{\mathcal{P}} f < \epsilon.$$

('Integrability means that upper and lower sums are arbitrarily close'.) □

Exercises
2. Determine whether f in examples 7–9 is integrable.
3. Let $\operatorname{dom} f = [0; 1] \times [0; 1]$ and $f(x, y) = x + y$ for each (x, y) in $\operatorname{dom} f$. Let $\mathscr{P}$ consist of n^2 equal square cells. Evaluate $^*\sum_{\mathscr{P}} f$ and $_*\sum_{\mathscr{P}} f$. Then evaluate $^*\int f$ and $_*\int f$.
4. Let $\operatorname{dom} f = [0; 1] \times [0; 1]$ and, for each (x, y) in $\operatorname{dom} f$,

$$f(x, y) = \begin{cases} 0 \text{ if } x \text{ or } y \text{ is irrational} \\ 1/n \text{ if } x \text{ is rational and } y = m/n \text{ where the} \\ \quad \text{fraction } m/n \text{ is in its lowest terms.} \end{cases}$$

Is f integrable?
5. Let $\operatorname{dom} f = [0; 1] \times [0; 1]$ and, for each (x, y) in $\operatorname{dom} f$

$$f(x, y) = \begin{cases} 0 \text{ if } x \text{ is irrational or zero} \\ 1/n \text{ if } x = m/n \text{ where the fraction} \\ \quad m/n \text{ is in its lowest terms.} \end{cases}$$

Is f integrable?
6. Let f be integrable. Let g have the same domain and be equal to f except at a finite number of points. Prove that g is integrable and $\int g = \int f$.
7. Let f be integrable. Show that $|f|$ is integrable and $\int |f| \geq |\int f|$.
8. Let f be integrable and $\mathbb{J}$ be a compact interval contained in $\operatorname{dom} f$. Prove that the restriction of f to $\mathbb{J}$ is integrable.

Theorem
Let f and g be integrable functions with the same domain. Then $f + g$ is integrable and

$$\int (f + g) = \int f + \int g.$$

Proof
Let ϵ be any positive number. Then there is a partition of the domain for which

$$\sum c(\mathbb{S}) \sup_{\mathbb{S}} f - \sum c(\mathbb{S}) \inf_{\mathbb{S}} f < \tfrac{1}{2}\epsilon \qquad (1)$$

and

$$\sum c(\mathbb{S}) \sup_{\mathbb{S}} g - \sum c(\mathbb{S}) \inf_{\mathbb{S}} g < \tfrac{1}{2}\epsilon \qquad (2)$$

where the summations are over all the cells $\mathbb{S}$ of the partition. For each cell,

$$\inf_{\mathbb{S}}(f + g) \geq \inf_{\mathbb{S}} f + \inf_{\mathbb{S}} g$$

and so

$$\sum c(\mathbb{S}) \inf_{\mathbb{S}}(f + g) \geq \sum c(\mathbb{S}) \inf_{\mathbb{S}} f + \sum c(\mathbb{S}) \inf_{\mathbb{S}} g. \qquad (3)$$

Similarly,

$$\sum c(\mathbb{S}) \sup_{\mathbb{S}}(f + g) \leq \sum c(\mathbb{S}) \sup_{\mathbb{S}} f + \sum c(\mathbb{S}) \sup_{\mathbb{S}} g. \quad (4)$$

Subtracting (3) from (4) and using (1) and (2), we obtain

$$\sum c(\mathbb{S}) \sup_{\mathbb{S}}(f + g) - \sum c(\mathbb{S}) \inf_{\mathbb{S}}(f + g) < \epsilon.$$

However, ϵ was any positive number. Therefore upper and lower sums for $f + g$ are arbitrarily close, and so $f + g$ is integrable.

Now $\int f + \int g$ lies between the right-hand sides of (3) and (4) which, by (1) and (2), are less than ϵ apart. Also, $\int (f + g)$ lies between the left-hand sides of (3) and (4) and therefore, by the inequalities themselves, between the right-hand sides. Therefore $\int f + \int g$ and $\int (f + g)$ are less than ϵ apart; this is true for every positive ϵ and so they must be equal. □

Theorem   If c is a number and f is an integrable function, then $\int cf$ exists and equals $c \int f$. (The proof is obvious.)   □

Theorem   Let $\mathbb{I}$ be a compact interval and let f and g be bounded functions with domain $\mathbb{I}$. If

$$f(\xi) \leq g(\xi) \qquad \text{for every } \xi \text{ in } \mathbb{I}$$

then

$$\overset{*}{\int} f \leq \overset{*}{\int} g$$

and

$$\int_{*} f \leq \int_{*} g.$$

The proof is obvious.   □

Corollary 1

If also f and g are integrable, then $\int f \leq \int g$.

Corollary 2

If f is integrable and its values are all non-negative, then

$$\int f \geq 0.$$

Corollary 3

If l is a lower bound for f and m is an upper bound, then

$$l \cdot c(\mathrm{dom}\, f) \le \int_* f \le m \cdot c(\mathrm{dom}\, f).$$

Proof

Compare the lower integral of f with that of the constant functions with the same domain and values l and m. $\quad\square$

    Corollary 3 also holds for upper integrals and, if f is integrable, for integrals.

    We have seen (pp. 105, 106) that, if f and g are integrable, so are $f + g$ and αf, where α is a number. From an algebraist's viewpoint, these results mean that integration is a linear operation on the set of integrable functions with a given domain.

    (Definition. Let $\mathcal{S}$ be a set of objects that can be added and can be multiplied by numbers. Then $\int$ is a **linear operator** on $\mathcal{S}$ if it is a function on $\mathcal{S}$ into $\mathbb{R}$ and if $\int (f + g) = \int f + \int g$ and $\int (\alpha f) = \alpha f$ for every f and g of $\mathcal{S}$ and every number α.)

    We might ask: What about $f - g$, fg, and f/g? It follows at once that $f - g$ is integrable if f and g are because $f - g = f + (-1)g$; it is true that fg is integrable but the proof is not straightforward, while f/g is not necessarily integrable even under the condition that g takes only non-zero values.

Lemma

If f is integrable, so is f^2.

Proof

Let c be a lower bound of f and $c + m$ an upper bound where $m > 0$. Let $h = f - c$. Then h is integrable and has lower bound zero and upper bound m. If ϵ is any positive number, there is a partition $\mathcal{P}$ for which

$$\overset{*}{\underset{\mathcal{P}}{\sum}} h - \underset{*}{\underset{\mathcal{P}}{\sum}} h \le \frac{\epsilon}{2m}. \tag{5}$$

For each cell $\mathbb{S}$ of $\mathcal{P}$,

$$\mathrm{sup}_{\mathbb{S}}(h^2) - \mathrm{inf}_{\mathbb{S}}(h^2) = (\mathrm{sup}_{\mathbb{S}}\, h)^2 - (\mathrm{inf}_{\mathbb{S}}\, h)^2$$

$$= (\mathrm{sup}_{\mathbb{S}}\, h - \mathrm{inf}_{\mathbb{S}}\, h)(\mathrm{sup}_{\mathbb{S}}\, h + \mathrm{inf}_{\mathbb{S}}\, h)$$

$$\le 2m(\mathrm{sup}_{\mathbb{S}} h - \mathrm{inf}_{\mathbb{S}}\, h). \tag{6}$$

Then

$$^*\sum_{\mathscr{P}} h^2 - {}_*\sum_{\mathscr{P}} h^2 = \sum_{\mathbb{S} \in \mathscr{P}} (\sup_{\mathbb{S}} h^2 - \inf_{\mathbb{S}} h^2) c(\mathbb{S})$$

$$\leq 2m \sum_{\mathbb{S} \in \mathscr{P}} (\sup_{\mathbb{S}} h - \inf_{\mathbb{S}} h) c(\mathbb{S}) \qquad \text{by (6)}$$

$$= 2m \left({}^*\sum_{\mathscr{P}} h - {}_*\sum_{\mathscr{P}} h \right)$$

$$\leq \epsilon. \qquad \text{by (5)}.$$

Therefore h^2 is integrable. But

$$f^2 = (h + c)^2$$
$$= h^2 + 2ch + c^2$$

and so f^2 is integrable. $\qquad \square$

Theorem If f and g have the same domain and are integrable, then fg is integrable.

Proof $fg = \frac{1}{4}(f + g)^2 - \frac{1}{4}(f - g)^2$, and $(f + g)^2$ and $(f - g)^2$ are integrable by the preceding lemma. $\qquad \square$

Problem 1. Let f be integrable and its domain be a square $\mathbb{I}$. For each positive integer n, let $\mathscr{P}_n$ be the partition of $\mathbb{I}$ consisting of n^2 equal squares. We define f_n by letting its domain be $\mathbb{I}$ and its value at any point of $\mathbb{I}$ be the supremum of f on the cell or cells containing the point. Each f_n is integrable. Does it follow that the sequence

$$\int f_1, \int f_2, \dots$$

converges to $\int f$?
What if f is bounded but not integrable?

By definition of an upper integral, there are upper sums arbitrarily close to it. The definition does not tell us how to find such an upper sum, but the following theorem does: we have only to make the mesh of the partition small enough.

Theorem If $\mathbb{I}$ is a compact interval and f is a bounded function whose domain is $\mathbb{I}$, and if ϵ is any positive number, then there is a positive number δ such that, whenever $\mathscr{P}$ is a partition of $\mathbb{I}$ whose mesh is less than δ,

$$^*\sum_{\mathscr{P}} f - \int^* f < \epsilon.$$

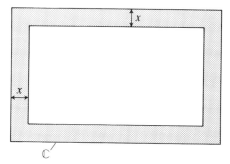

Fig. 5.2

Proof

Let $\mathcal{Q}$ be a partition of $\mathbb{I}$ such that

$$^*\!\sum_{\mathcal{Q}} f < ^*\!\!\int f + \epsilon/2.$$

Let $m > \sup f - \inf f$. For each cell $\mathbb{C}$ of $\mathcal{Q}$ and each number x less than half the smallest side of any cell, let $\mathbb{C}_x$ be the set of points represented diagrammatically by the hollow rectangle shown in Fig. 5.2. There is, clearly, a number δ such that

$$\sum c(\mathbb{C}_\delta) < \epsilon/2m,$$

where the sum is over all the cells of $\mathcal{Q}$.

Let $\mathcal{P}$ be any partition of $\mathbb{I}$ whose mesh is less than δ and let $\mathcal{R}$ consist of the intersections of the cells of $\mathcal{P}$ with those of $\mathcal{Q}$. Let us divide the cells of $\mathcal{P}$ into two classes. If a cell of $\mathcal{P}$ is entirely contained in a cell of $\mathcal{Q}$, it is in class I; if not, it is in class II. Cells of class I are also cells of $\mathcal{R}$, and their contribution to

$$^*\!\sum_{\mathcal{P}} f - ^*\!\sum_{\mathcal{R}} f \tag{7}$$

is zero.

A cell $\mathbb{B}$ of class II is the union of cells $\mathbb{A}$ of $\mathcal{R}$, and its contribution to (7) is

$$\sup_{\mathbb{B}} f \cdot c(\mathbb{B}) - \sum_{\mathbb{A} \subseteq \mathbb{B}} \sup_{\mathbb{A}} f \cdot c(\mathbb{A})$$

$$= \sum_{\mathbb{A} \subseteq \mathbb{B}} (\sup_{\mathbb{B}} f - \sup_{\mathbb{A}} f) c(\mathbb{A})$$

$$\leq \sum_{\mathbb{A} \subseteq \mathbb{B}} m c(\mathbb{A})$$

$$= m c(\mathbb{B}).$$

Thus

$$^*\sum_{\mathscr{P}} f - {}^*\sum_{\mathscr{R}} f \le m \sum c(\mathbb{B}),$$

where the sum on the right-hand side is over all cells $\mathbb{B}$ of class II. But each cell of class II crosses a boundary of a cell $\mathbb{C}$ of $\mathscr{Q}$, and so is contained in the union of the $\mathbb{C}_\delta$ because the mesh of $\mathscr{P}$ is less than δ. Therefore

$$\sum c(\mathbb{B}) < \epsilon/2m$$

and so

$$^*\sum_{\mathscr{P}} f - {}^*\sum_{\mathscr{R}} f < \epsilon/2.$$

Then

$$^*\sum_{\mathscr{P}} f < {}^*\sum_{\mathscr{R}} f + \epsilon/2$$

$$\le {}^*\sum_{\mathscr{Q}} f + \epsilon/2$$

$$< {}^*\!\int f + \epsilon.$$

Iterated integration

We now turn to a very practical result, a theorem that reduces the evaluation of an integral in $\mathbb{R}^2$ to the evaluation of two integrals in $\mathbb{R}$.

The inspiration behind it is fairly obvious. If f is a positive-valued continuous function with a compact interval in $\mathbb{R}$ for domain, the set

$$\{(x, y) : 0 \le y \le f(x), x \in \operatorname{dom} f\}$$

can be represented diagrammatically as a two-dimensional region, the 'region under the graph' (Fig. 5.3(a)). The area of this region is $\int f$. Indeed, the main motive behind the definition of integral is to make this so.

In the same way, if f is a positive-valued continuous function with a compact interval in $\mathbb{R}^2$ for domain, the set

$$\{(x, y, z) : 0 \le z \le f(x, y), (x, y) \in \operatorname{dom} f\}$$

can be represented diagrammatically as a three-dimensional region, the 'region under the graph' (Fig. 5.3(b)). The volume of this region is $\int f$. If we slice this region by a plane perpendicular to the x axis we obtain a cross-section of the

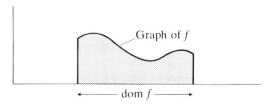

(a) $\square = \{(\zeta,\eta): 0 \leqslant \eta \leqslant f(\zeta), \zeta \in \text{dom } f\}$

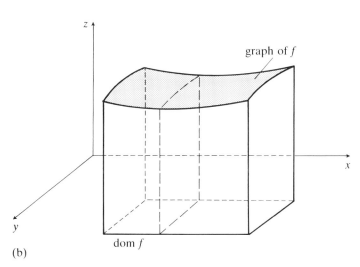

(b)

Fig. 5.3

region. Let the area of the cross-section of the region by the plane $x = c$ be $u(c)$ for each c for which the plane does cut the region. It is clear, for geometrical reasons, that the volume is the integral of the area of cross-section. But the area of cross-section is itself an integral:

$$u(c) = \int f(c, \cdot).$$

To sum up,

$$\int f = \int u \quad \text{where} \quad u(c) = \int f(c, \cdot).$$

Thus $\int f$, which is an integral in $\mathbb{R}^2$, can be obtained by evaluating $\int (c, \cdot)$ and $\int u$, each of which is an integral in $\mathbb{R}$.

We prove the theorem, of course, without appealing to geometry, and we do not confine ourselves to positive-valued functions.

Theorem Let f be integrable and u be defined by

$$u(x) = \int_* f(x, \cdot).$$

Then u is integrable and

$$\int f = \int u.$$

Proof Let $\mathcal{P}$ be a partition of $\operatorname{dom} f$. Then $\mathcal{P} = \mathcal{Q} \times \mathcal{R}$ where $\operatorname{dom} f = \mathbb{I} \times \mathbb{J}$, and $\mathcal{Q}$ is a partition of $\mathbb{I}$ and $\mathcal{R}$ of $\mathbb{J}$. Each cell $\mathbb{S}$ of $\mathcal{P}$ is of the form $\mathbb{T} \times \mathbb{U}$ where $\mathbb{T} \in \mathcal{Q}$ and $\mathbb{U} \in \mathcal{R}$. Then

$$c(\mathbb{S}) \inf_{\mathbb{S}} f = c(\mathbb{T}) c(\mathbb{U}) \inf_{\mathbb{T} \times \mathbb{U}} f$$

and so

$$_*\sum_{\mathcal{P}} f = \sum_{\mathbb{T} \in \mathcal{Q}} c(\mathbb{T}) \sum_{\mathbb{U} \in \mathcal{R}} c(\mathbb{U}) \inf_{\mathbb{T} \times \mathbb{U}} f. \qquad (8)$$

If $x \in \mathbb{T}$, then

$$\inf_{\mathbb{T} \times \mathbb{U}} f \leq \inf_{\mathbb{U}} f(x, \cdot)$$

and then

$$\sum_{\mathbb{U} \in \mathcal{R}} c(\mathbb{U}) \inf_{\mathbb{T} \times \mathbb{U}} f \leq \sum_{\mathbb{U} \in \mathcal{R}} c(\mathbb{U}) \inf_{\mathbb{U}} f(x, \cdot)$$

$$\leq \int_* f(x, \cdot)$$

$$= u(x).$$

Because this holds for each x of $\mathbb{T}$,

$$\sum_{\mathbb{U} \in \mathcal{R}} c(\mathbb{U}) \inf_{\mathbb{T} \times \mathbb{U}} f \leq \inf_{\mathbb{T}} u.$$

Therefore, by (8),

$$_*\sum_{\mathcal{P}} f \leq \sum_{\mathbb{T} \in \mathcal{Q}} c(\mathbb{T}) \inf_{\mathbb{T}} u$$

$$= {}_*\sum_{\mathcal{Q}} u,$$

and so

$$\int f \leq \int_* u.$$

If $v(x) = {}^*\!\int f(x, \cdot)$ for each x, we have

$$\int f \geq \int^* v \quad \text{(similarly)}$$

$$\geq \int^* u \geq \int_* u.$$

Hence all these integrals are equal; in particular

$$\int^* u = \int_* u = \int f.$$

Corollary 1

$$\int v = \int f.$$

Corollary 2

Under the same conditions, $\int f = \int p = \int q$, where

$$p(y) = \int_* f(\cdot, y) \quad \text{and} \quad q(y) = \int^* f(\cdot, y).$$

Corollary 3

If f is continuous, then so are $f(x, \cdot)$ and $f(\cdot, y)$ and so these two functions are integrable. In this case we can simply define u etc. as integrals and omit all mention of upper and lower integrals.

Example   10. Let f have domain $[0; 1] \times [0; 1]$ and let

$$f(x, y) = x + y \text{ for each } (x, y) \text{ in the domain.}$$

With the notation of the theorem (or, rather, corollary 3 because f is continuous)

$$u(x) = \int f \quad \text{where} \quad f(y) = x + y \text{ for } 0 \leq y \leq 1$$

$$= x + 1/2.$$

Then

$$\int f = \int u$$

$$= 1.$$

Exercise   9. Let f have domain $[-1; 1] \times [0; 2]$ and

$$f(x, y) = x^2 y$$

for each (x, y) in the domain. Evaluate $\int f$.

Problem   2. Let f have domain $[0; 1] \times [0; 1]$ and

$$f(x, y) = \begin{cases} 0 \text{ if } x \text{ or } y \text{ is irrational} \\ 1/n \text{ if } x \text{ is rational and } y = m/n, \\ \quad \text{where the fraction } m/n \text{ is in its} \\ \quad \text{lowest terms.} \end{cases}$$

(We know from exercise 4 that f is integrable.) Apply the theorem or its corollaries to this f.

Integrability

What functions are integrable? In one dimension, a bounded function which is continuous except at a finite number of points is integrable. Informally, a function which is continuous except on a negligibly small set is integrable. A similar result is true in two dimensions. We start by making clear what sets we consider to be negligibly small.

Definition   A subset A of $\mathbb{R}^2$ is **negligible** if, for each positive ϵ, it can be covered by a finite set of compact intervals whose contents sum to less than ϵ.   □

Examples   11. $\{(x, x) : 0 \leq x \leq 1\}$ is negligible.

Proof   Given any positive ϵ, choose an integer n greater than $1/\epsilon$ and let $\mathscr{P}$ be the partition of $[0; 1] \times [0; 1]$ into n^2 equal squares. The n squares along a diagonal cover the given set, and their contents sum to $1/n$.   □

12. If $a < b$ and $c < d$, then $[a; b] \times [c; d]$ is not negligible. In fact, the contents of any set of compact intervals that cover it must sum to at least $(b - a)(d - c)$. Similarly, the set of rational points in $[a; b] \times [c; d]$ is not negligible.

Clearly, any subset of a negligible set is negligible, and so is the union of a finite number of negligible sets. Every negligible set is bounded.

In fact, a set $\mathbb{A}$ is negligible if and only if, for each positive ϵ and each compact interval $\mathbb{I}$ containing $\mathbb{A}$, there is a partition $\mathscr{P}$ of $\mathbb{I}$ such that the contents of the cells of $\mathscr{P}$ that intersect $\mathbb{A}$ sum to less than ϵ. (Proof. If the partition exists, the cells that intersect $\mathbb{A}$ form an ϵ-covering. Conversely, if $\mathbb{A}$ is negligible, take an ϵ-covering and extend all sides of the covering intervals to the edges of $\mathbb{I}$ to form $\mathscr{P}$.)

Exercise   10. Is $\{(1/n, y):0\leq y \leq 1 \text{ and } n \text{ is a positive integer}\}$ negligible?

Problem   3. Show that the boundary of a negligible set is negligible.

Theorem   Let f be a bounded function whose domain is a compact interval $\mathbb{I}$, and

$$\{\xi \in \mathbb{I} : f \text{ is not continuous at } \xi\} \tag{9}$$

be negligible. Then f is integrable. ('A function that is continuous nearly everywhere is integrable'.)

Proof   Let ϵ be any positive number. Let m be an upper bound of $|f|$. There is a finite set of compact intervals $\mathbb{J}_1 \ldots \mathbb{J}_h$ which cover the set (9) for which

$$\sum_{i=1}^{h} c(\mathbb{J}_i) < \frac{\epsilon}{4m}.$$

Let $\mathbb{K}$ be the set of points of $\mathbb{I}$ not in the interior of any $\mathbb{J}_i$. $\mathbb{K}$ is closed and bounded and therefore compact. Consequently f is uniformly continuous on $\mathbb{K}$ and so there is a positive δ such that

$$|f(\xi) - f(\eta)| < \frac{\epsilon}{2c(\mathbb{I})}$$

whenever

$$\xi \in \mathbb{K}, \ \eta \in \mathbb{K}, \text{ and } |\xi - \eta| < \delta.$$

Let $\mathscr{P}$ be a partition of $\mathbb{I}$ whose mesh is less than $\frac{1}{2}\delta$ and no cell of which intersects more than one of the $\mathbb{J}_i$. Let $\mathscr{Q}$ be the set of cells that do intersect the $\mathbb{J}_i$. Then

$$\sum_{\mathbb{S} \in \mathscr{Q}} c(\mathbb{S}) = \sum_{i=1}^{h} c(\mathbb{J}_i)$$

$$< \epsilon / 4m$$

and so

$$\sum_{\mathbb{S} \in \mathscr{Q}} (\sup_{\mathbb{S}} f - \inf_{\mathbb{S}} f) \cdot c(\mathbb{S}) < \frac{2m\epsilon}{4m}$$

$$= \tfrac{1}{2}\epsilon. \tag{10}$$

Each of the other cells is contained in $\mathbb{K}$. If ξ and η belong to one such cell $\mathbb{S}$, then

$$\xi \in \mathbb{K}, \quad \eta \in \mathbb{K}, \quad \text{and} \quad |\xi - \eta| < \delta$$

and so

$$|f(\xi) - f(\eta)| < \frac{\epsilon}{2c(\mathbb{I})}.$$

Consequently

$$\sup_{\mathbb{S}} f - \inf_{\mathbb{S}} f \le \frac{\epsilon}{2c(\mathbb{I})}$$

and so

$$\sum_{\mathbb{S} \notin \mathscr{Q}} (\sup_{\mathbb{S}} f - \inf_{\mathbb{S}} f) \cdot c(\mathbb{S}) \le \sum_{\mathbb{S} \notin \mathscr{Q}} c(\mathbb{S}) \cdot \frac{\epsilon}{2c(\mathbb{I})}$$

$$\le c(\mathbb{I}) \cdot \frac{\epsilon}{2c(\mathbb{I})}$$

$$= \tfrac{1}{2}\epsilon. \tag{11}$$

From (10) and (11)

$$\sum (\sup_{\mathbb{S}} f - \inf_{\mathbb{S}} f) \cdot c(\mathbb{S}) < \epsilon$$

where the sum is over all cells of $\mathscr{P}$. Thus upper and lower sums are arbitrarily close and so f is integrable. $\square$

This is not the most powerful theorem that we could have proved. Let us call a set 'Lebesgue-negligible' if, for each positive ϵ, we can cover it by an infinite sequence of compact intervals whose contents sum to less than ϵ. Every negligible set is Lebesgue-negligible but not vice versa. (For example, the set of rational points in $\mathbb{R}^2$ is Lebesgue-negligible because we can arrange them as a sequence and cover the nth point by an interval of content less than $\epsilon/2^n$.)

The reason for the name is that the French mathematician Henri Lebesgue developed a more sophisticated theory of measure and of integration than the one we are presenting, one difference being that Lebesgue allowed infinite sequences in his definitions; our 'Lebesgue-negligible' sets are the sets with zero measure in his theory.

The more powerful theorem is as follows.

Theorem   A bounded function f whose domain is a compact interval is integrable if and only if

$$\{\xi \in \mathrm{dom}\, f : f \text{ is not continuous at } \xi\}$$

is Lebesgue-negligible.

A proof can be found in *Calculus on manifolds* by M. Spivak (1965) and in other texts.   □

Theorem   If $\mathbb{I}$ is a compact interval, if $\mathbb{C} \subseteq \mathbb{I}$, and if f is the function with domain $\mathbb{I}$ for which

$$f(\xi) = \begin{cases} 1 & \text{if } \xi \in \mathbb{C} \\ 0 & \text{if } \xi \notin \mathbb{C} \end{cases}$$

then f is integrable if and only if bdy $\mathbb{C}$ is negligible.

Proof f is continuous on $\mathbb{I}$ except at points of bdy $\mathbb{C}$. Then if bdy $\mathbb{C}$ is negligible, f is integrable by the theorem on p. 115.

Conversely, if f is integrable, then for any positive ϵ there is a partition $\mathscr{P}$ of $\mathbb{I}$ such that

$$\sum_{\mathbb{S} \in \mathscr{P}} (\sup_\mathbb{S} f - \inf_\mathbb{S} f) \cdot c(\mathbb{S}) < \epsilon.$$

Let $\mathscr{Q}$ be the set of cells of $\mathscr{P}$ that intersect bdy $\mathbb{C}$: they cover bdy $\mathbb{C}$. If $\mathbb{S} \in \mathscr{Q}$, then $\mathbb{S}$ contains both a point in $\mathbb{C}$ and a point not in $\mathbb{C}$ and so $\inf_\mathbb{S} f = 0$ and $\sup_\mathbb{S} f = 1$. Therefore

$$\sum_{\mathbb{S} \in \mathscr{Q}} c(\mathbb{S}) = \sum_{\mathbb{S} \in \mathscr{Q}} (\sup_\mathbb{S} f - \inf_\mathbb{S} f) \cdot c(\mathbb{S})$$

$$\leq \sum_{\mathbb{S} \in \mathscr{P}} (\sup_\mathbb{S} f - \inf_\mathbb{S} f) \cdot c(\mathbb{S})$$

$$< \epsilon.\qquad\qquad\qquad\qquad\qquad □$$

Problems   4. If f is integrable, does it follow that $f(x, \cdot)$ is integrable?
5. $\int f = 0$. $f(\xi) \geq 0$ for every ξ in dom f. Does it follow that $\{\xi : f(\xi) > 0\}$ is negligible?

Integration over bounded sets

Definitions

Let f be a function in $\mathbb{R}^2$ into $\mathbb{R}$ and let $\mathbb{A} \subseteq \mathrm{dom}\, f$. We define a function $f_\mathbb{A}|$ as follows:

$$f_\mathbb{A}|(\xi) = \begin{cases} f(\xi) & \text{if } \xi \in \mathbb{A} \\ 0 & \text{if } \xi \notin \mathbb{A} \text{ but } \xi \in \mathbb{R}^2. \end{cases}$$

Then dom $f_{\mathbb{A}}|$ is the whole of $\mathbb{R}^2$. Informally '$f_{\mathbb{A}}|$ behaves like f on $\mathbb{A}$ and is zero outside'. It is not quite the same as $f_{\mathbb{A}}$, the restriction of f to $\mathbb{A}$, whose domain is $\mathbb{A}$. □

Lemma Let f be a function in $\mathbb{R}^2$ into $\mathbb{R}$, let $\mathbb{A} \subseteq \mathrm{dom}\, f$, and let $\mathbb{I}$ and $\mathbb{J}$ be two compact intervals, each containing $\mathbb{A}$. Then

$$\int f_{\mathbb{A}}|_{\mathbb{I}} = \int f_{\mathbb{A}}|_{\mathbb{J}}$$

if either integral exists.

The proof is obvious. □

Definition Let f be a function in $\mathbb{R}^2$ into $\mathbb{R}$, and let $\mathbb{A} \subseteq \mathrm{dom}\, f$. If there is a compact interval $\mathbb{I}$ containing $\mathbb{A}$ such that

$$\int f_{\mathbb{A}}|_{\mathbb{I}}$$

exists, we say that f is **integrable on** $\mathbb{A}$ and we denote this integral (which is, by the lemma, independent of the choice of $\mathbb{I}$) by

$$\int_{\mathbb{A}} f.$$

Whether or not $\int f_{\mathbb{A}}|_{\mathbb{I}}$ exists, if f is bounded on $\mathbb{A}$ we can define

$$\int_{*\mathbb{A}} f \quad \text{to be} \quad \int_{*} f_{\mathbb{A}}|_{\mathbb{I}} \qquad \text{and} \qquad \overset{*}{\int_{\mathbb{A}}} f \quad \text{to be} \quad \overset{*}{\int} f_{\mathbb{A}}|_{\mathbb{I}}.$$

(As above, they are independent of the choice of $\mathbb{I}$.)

Clearly if $f = g$ on $\mathbb{A}$, then, no matter how differently f and g behave outside $\mathbb{A}$,

$$\int_{\mathbb{A}} f = \int_{\mathbb{A}} g$$

if either integral exists. (The same applies to upper and lower integrals.)

Exercises 11. Show that if f and g are integrable on $\mathbb{A}$, then so are fg, $f + g$, $|f|$, and κf, where κ is a real number, and that

$$\int_{\mathbb{A}} (f + g) = \int_{\mathbb{A}} f + \int_{\mathbb{A}} g$$

$$\int_{\mathbb{A}} \kappa f = \kappa \int_{\mathbb{A}} f$$

$$\int_{\mathbb{A}} |f| \geq \left| \int_{\mathbb{A}} f \right|.$$

12. Show that if $\int_{\mathbb{A}} f$ and $\int_{\mathbb{A}} g$ exist and $f(\xi) \le g(\xi)$ for every ξ of $\mathbb{A}$, then

$$\int_{\mathbb{A}} f \le \int_{\mathbb{A}} g.$$

13. Show that if $\int_{\mathbb{A}} f$ exists and if, for each x,

$$u(x) = \int_{*\mathbb{A}_x} f(x, \cdot),$$

where

$$\mathbb{A}_x = \{y : (x, y) \in \mathbb{A}\},$$

then u is integrable on $\mathbb{B}$, where

$$\mathbb{B} = \{x : \exists y \text{ such that } (x, y) \in \mathbb{A}\},$$

and that

$$\int_{\mathbb{A}} f = \int_{\mathbb{B}} u.$$

14. Let $\mathbb{A} = (0; 1) \times (0; 1)$ and $f(\xi) = 1$ for every ξ in $\mathbb{R}^2$. Is f integrable on $\mathbb{A}$ and, if so, what is $\int_{\mathbb{A}} f$? If not, what are $^*\int_{\mathbb{A}} f$ and $_*\int_{\mathbb{A}} f$?

15. If f is integrable on $\mathbb{A}$ and on $\mathbb{B}$, is it necessarily integrable on $\mathbb{A} \cap \mathbb{B}$ or on $\mathbb{A} \cup \mathbb{B}$?

16. If the set of points of $\mathbb{A}$ at which f is discontinuous is negligible, and if $\mathbb{A}$ and f are bounded, does $\int_{\mathbb{A}} f$ necessarily exist?

Problems

6. Given that f is non-negative on $\mathbb{A}$ and that $\int_{\mathbb{A}} f = 0$, does it follow that the set of points of $\mathbb{A}$ at which f is positive is negligible?

7. If $\int_{\mathbb{A} \times \mathbb{B}} f$ exists (where $\mathbb{A} \subseteq \mathbb{R}$ and $\mathbb{B} \subseteq \mathbb{R}$) and $g(x, y) = f(y, x)$ for every (x, y) in $\mathbb{B} \times \mathbb{A}$, does it follow that

$$\int_{\mathbb{B} \times \mathbb{A}} g = \int_{\mathbb{A} \times \mathbb{B}} f?$$

8. If $\int_{\mathbb{A}} f$ exists, where $\mathbb{A}$ is a compact interval, if $g(x, y) = f(cx, dy)$, and if $\mathbb{B} = \{(x, y) : (cx, dy) \in \mathbb{A}\}$, does it follow that $\int_{\mathbb{B}} g$ exists? If so, what is the relation between the two integrals?

What if $\mathbb{A}$ is a more general subset of $\mathbb{R}^2$, not necessarily an interval?

If f is bounded on $\mathbb{A}$ but not necessarily integrable on $\mathbb{A}$, what can be said about the upper and lower integrals?

9. If $\int_{\mathbb{A}} f$ exists, where $\mathbb{A}$ is a compact interval in $\mathbb{R}^2$, if $g(x, y) = f(x + y, y)$, and if

$$\mathbb{B} = \{(x, y) : (x + y, y) \in \mathbb{A}\},$$

does it follow that $\int_{\mathbb{B}} g$ exists? If so, what is the relation between the two integrals?

10. Can you generalize Problems 8 and 9 any further? What about $\int_{\mathbb{B}} f \circ g$ where $\mathbb{B} = g^{-}\mathbb{A}$ and g is any linear function on $\mathbb{R}^2$ onto $\mathbb{R}^2$? (Reminder. $g^{-}\mathbb{A}$ is the inverse image of $\mathbb{A}$ under g.)

11. $\mathbb{A} \subseteq \mathbb{R}^2$ and $\mathbb{B}$ is the union of $\mathbb{A}$ and its boundary. $\int_{\mathbb{A}} f$ and $\int_{\mathbb{B}} f$ both exist. Need they necessarily be equal?

We can now extend our definition of content of sets in $\mathbb{R}^2$; so far it applies only to compact intervals.

Definitions

If $\mathbb{I}$ is a non-null compact interval,

$$\int_{\mathbb{I}} \mathbf{1} = c(\mathbb{I})$$

(where $\mathbf{1}$ denotes the constant function with domain $\mathbb{R}^2$ and value 1). This fact inspires us to define the **content** $c(\mathbb{A})$ of any subset $\mathbb{A}$ of $\mathbb{R}^2$ to be $\int_{\mathbb{A}} \mathbf{1}$, provided that this integral exists. If it does not, then $\mathbb{A}$ does not have a content. We also define the **upper content** $^*c(\mathbb{A})$ to be $= {}^*\!\int_{\mathbb{A}} \mathbf{1}$, and the **lower content** similarly. □

The second theorem on p. 117 tells us that a bounded set has content if and only if its boundary is negligible. It is clear (see exercise 14) that $(a; b) \times (c; d)$ has the same content as $[a; b] \times [c; d]$.

We could develop the theory of content in its own right, but we shall confine ourselves to the results we need for integration. The first of these is as follows.

Theorem

If $c(\mathbb{F})$ and $c(\mathbb{G})$ exist and $\mathbb{F} \cap \mathbb{G} = \emptyset$, then $c(\mathbb{F} \cup \mathbb{G})$ exists and

$$c(\mathbb{F} \cup \mathbb{G}) = c(\mathbb{F}) + c(\mathbb{G}).$$

Proof

$\mathbb{F}$ and $\mathbb{G}$ must be bounded. Let $\mathbb{I}$ be a compact interval containing both. Let f and g be functions with domain $\mathbb{I}$, and let f have value 1 on $\mathbb{F}$ and 0 off $\mathbb{F}$, and g have value 1 on $\mathbb{G}$ and 0 off $\mathbb{G}$. Then $f + g$ has value 1 on $\mathbb{F} \cup \mathbb{G}$ and 0 off $\mathbb{F} \cup \mathbb{G}$. Therefore

$$c(\mathbb{F}) + c(\mathbb{G}) = \int f + \int g$$

$$= \int (f + g)$$

$$= c(\mathbb{F} \cup \mathbb{G}). \qquad \square$$

Remark A set is negligible if and only if it has zero content.

Proof Let $\mathbb{A}$ be a negligible set. It must be bounded: let $\mathbb{I}$ be a compact interval containing it. Let f be the function with domain $\mathbb{I}$ for which

$$f(\xi) = \begin{cases} 1 & \text{if } \xi \in \mathbb{A} \\ 0 & \text{if } \xi \notin \mathbb{A}. \end{cases}$$

If ϵ is any positive number, cover $\mathbb{A}$ with a finite set $\mathscr{R}$ of compact intervals whose contents sum to less than ϵ and let $\mathscr{P}$ be a partition of $\mathbb{I}$ none of whose cells overlaps more than one interval of $\mathscr{R}$. Let $\mathscr{Q}$ be the set of those cells $\mathbb{S}$ of $\mathscr{P}$ that intersect $\mathbb{A}$.

If $\mathbb{S} \notin \mathscr{Q}$, then $\sup_{\mathbb{S}} f = 0$. In any case

$$0 \le \sup_{\mathbb{S}} f \le 1.$$

Therefore

$$0 \le \sum_{\mathbb{S} \in \mathscr{P}} \sup_{\mathbb{S}} f \cdot c(\mathbb{S}) \le \sum_{\mathbb{S} \in \mathscr{Q}} c(\mathbb{S}) < \epsilon$$

because each cell $\mathbb{S}$ in $\mathscr{Q}$ is contained in one of the intervals of $\mathscr{R}$. Then also

$$0 \le \sum_{\mathbb{S} \in \mathscr{P}} \inf_{\mathbb{S}} f \cdot c(\mathbb{S}) \le \epsilon.$$

Therefore

$$\int f = 0$$

and so

$$c(\mathbb{A}) = 0.$$

Conversely, let $\mathbb{A}$ have content zero. Then there exist $\mathbb{I}$ and f as above, and

$$\int f = 0.$$

Therefore if ϵ is any positive number there is a partition $\mathscr{P}$ of $\mathbb{I}$ such that

$$\sum_{\mathbb{S} \in \mathscr{P}} (\sup_{\mathbb{S}} f - \inf_{\mathbb{S}} f) \cdot c(\mathbb{S}) < \epsilon.$$

If $\mathbb{S}$ does not intersect $\mathbb{A}$, $\sup_{\mathbb{S}} f = \inf_{\mathbb{S}} f = 0$ and so, if we sum over those cells that do intersect $\mathbb{A}$,

$$\sum (\sup_{\mathbb{S}} f - \inf_{\mathbb{S}} f) \cdot c(\mathbb{S}) < \epsilon.$$

But every $\mathbb{S}$ contains points not in $\mathbb{A}$ because a negligible set cannot contain a compact interval (except in the trivial case where the interval has zero content) and so if $\mathbb{S}$ intersects $\mathbb{A}$, $\sup_{\mathbb{S}} f - \inf_{\mathbb{S}} f = 1$. Therefore

$$\sum c(\mathbb{S}) < \epsilon$$

and these cells form a covering of $\mathbb{A}$ by a finite number of compact intervals. $\qquad\square$

From now on we shall abandon the term 'negligible' and replace it by 'of content zero'.

Theorem If f is integrable on $\mathbb{G}$, $\mathbb{A} \subseteq \mathbb{G}$, and $c(\mathrm{bdy}\,\mathbb{A}) = 0$, then f is integrable on $\mathbb{A}$.

Proof Let $\mathbb{I}$ be a compact interval containing $\mathbb{G}$. Let g be the function with domain $\mathbb{I}$ for which

$$g(\xi) = \begin{cases} 1 & \text{if } \xi \in \mathbb{A} \\ 0 & \text{if } \xi \notin \mathbb{A}. \end{cases}$$

Then g is continuous on $\mathbb{I}$ except at points of bdy $\mathbb{A}$ and so is integrable.

$$f_{\mathbb{A}}|_{\mathbb{I}} = f_{\mathbb{G}}|_{\mathbb{I}} \cdot g$$

and therefore $f_{\mathbb{A}}|_{\mathbb{I}}$ is integrable; that is to say, f is integrable on $\mathbb{A}$.

Exercises

17. Find, from the definition, the content of the set of pairs which, in the usual diagrammatic representation, occupies the triangle with vertices at the points represented by $(0,0)$, (a, b), and $(c, 0)$ where $a > 0$, $b > 0$, and $c > a$.

18. Prove that

$$\{(x, y) : x^2 + y^2 \le 1\}$$

has content. (We call this content π.)

19. Find the content of

$$\{(x, y) : x^2 + y^2 < 1\}$$

and of

$$\{\xi \in \mathbb{R}^2 : |\xi - \alpha| \le r\}.$$

20. Why is it obvious that if $\mathrm{dom}\, f \supseteq \mathbb{A}$, if $\mathbb{A}$ has content, and if f is constant, then

$$\int_{\mathbb{A}} f = c(\mathbb{A}) \times \text{value of } f?$$

21. Find the content of the set of pairs which, in the usual diagrammatic representation, occupies the parallelogram with vertices at α, $\alpha + \kappa$, $\alpha + \lambda$, and $\alpha + \kappa + \lambda$. Express the result as a determinant.

22. Why is it clear that, if $\mathbb{A}$ is bounded and $\mathbb{B} \subseteq \mathbb{A}$, then $^*c(\mathbb{B}) \leq {}^*c(\mathbb{A})$ and $_*c(\mathbb{B}) \leq {}_*c(\mathbb{A})$? What can we say about the relation between $c(\mathbb{B})$ and $c(\mathbb{A})$?

Problems 12. If h is a continuous function in $\mathbb{R}^2$ into $\mathbb{R}^2$ whose domain contains $\mathbb{A}$, where $\mathbb{A} = [0; 1] \times \{0\}$, does it follow that $c(h(\mathbb{A})) = 0$?

13. If $\mathbb{D}$ has content zero and $\text{dom} f \supseteq \mathbb{D}$, does it follow that $\int_{\mathbb{D}} f$ exists and is zero?

14. Let $r_1, r_2, \ldots$ be the rational numbers between 0 and 1, arranged as a sequence in the usual order, namely $1/2$, $1/3$, $2/3$, $1/4$, $3/4$, $1/5$, $2/5$, $3/5$, $4/5$, $1/6$, $5/6$, $1/7$, $2/7, \ldots$. Let $\mathbb{S}_n$ be $[0; 1] \times [r_n - 1/10^n; r_n + 1/10^n]$. Let $\mathbb{S}$ be the union of all the $\mathbb{S}_n$. Is it true or false that

$$c(\mathbb{S}) \leq \sum_{n=1}^{\infty} c(\mathbb{S}_n)$$

$$= \sum_{n=1}^{\infty} 2/10^n$$

$$= 2/9?$$

15. If a set has content, must its interior and its closure have content? (The closure of a set is the union of the set with its boundary.)

We now prove some lemmas that we shall need later.

Lemma 1 If $\mathbb{Q} \subseteq \text{dom} h \subseteq \mathbb{R}^2$ and h is into $\mathbb{R}^2$ and has an inverse continuous at each point of $\text{bdy}[h(\mathbb{Q})]$, then

$$\text{bdy}[h(\mathbb{Q})] \subseteq h(\text{bdy } \mathbb{Q}).$$

Proof Let α be any point of $\text{bdy}[h(\mathbb{Q})]$ and $\mathbb{U}$ be any neighbourhood of $h^-(\alpha)$, where h^- is the inverse of h. Because h^- is continuous at α, there is a neighbourhood $\mathbb{V}$ of α such that

$$h^-(\xi) \in \mathbb{U} \quad \text{whenever} \quad \xi \in \mathbb{V}. \tag{12}$$

Because $\alpha \in \text{bdy}[h(\mathbb{Q})]$, $\mathbb{V}$ contains a point η in $h(\mathbb{Q})$ and a point ζ not in $h(\mathbb{Q})$. Then, by (12),

$$h^-(\eta) \in \mathbb{U} \quad \text{and} \quad h^-(\zeta) \in \mathbb{U}$$

and clearly

$$h^-(\eta) \in \mathbb{Q} \quad \text{and} \quad h^-(\zeta) \notin \mathbb{Q}.$$

Thus $\mathbb{U}$ contains a point in $\mathbb{Q}$ and a point not in $\mathbb{Q}$. But $\mathbb{U}$ was any neighbourhood of $h^-(\alpha)$, and so

$$h^-(\alpha) \in \text{bdy } \mathbb{Q}.$$

Therefore

$$\alpha \in h(\text{bdy } \mathbb{Q}). \qquad \square$$

Although this lemma is all that we shall actually use, we might as well state a symmetrical theorem that can be proved immediately from it.

Theorem   If $\mathbb{Q} \subseteq \text{dom } h \subseteq \mathbb{R}^2$ and h is into $\mathbb{R}^2$ and is continuous at each point of bdy $\mathbb{Q}$ and has an inverse continuous at each point of $\text{bdy}[h(\mathbb{Q})]$, then

$$\text{bdy}[h(\mathbb{Q})] = h(\text{bdy } \mathbb{Q}). \qquad \square$$

Lemma 2   If $\mathbb{L} = \{(a, x) : b \leq x \leq b + d\}$ and h is in $\mathbb{R}^2$ into $\mathbb{R}^2$ and is continuously differentiable at each point of $\mathbb{L}$, then $h(\mathbb{L})$ has content zero.

Proof $\mathbb{L}$ is compact; therefore each $D_2 h_j$ is bounded on $\mathbb{L}$. Let

$$m \geq \sup_{\mathbb{L}} |D_2 h_j| \text{ for each } j.$$

Let ϵ be any positive number and n be an integer greater than $4d^2 m^2 / \epsilon$. Then, whenever (a, x) and (a, y) belong to $\mathbb{L}$,

$$|h(a, x) - h(a, y)|$$

$$= |(x - y) \cdot D_2 h(a, z)| \qquad \text{for some } z \text{ between } x \text{ and } y$$

$$\leq |x - y| \cdot m.$$

For each integer i from 0 to n, set

$$x_i = b + \frac{id}{n}$$

so that $x_0 = b$ and $x_n = b + d$. If (a, x) is between (a, x_i) and (a, x_{i+1}), then

$$|h(a, x) - h(a, x_i)| \leq |x - x_i| \cdot m$$

$$\leq \frac{dm}{n}.$$

Then the image under h of the segment of $\mathbb{L}$ from (a, x_i) to (a, x_{i+1}) lies in a square interval of side $2dm/n$. Thus $h(\mathbb{L})$ is covered by square intervals whose contents sum to $4d^2 m^2 / n$, which is less than ϵ. $\qquad \square$

Lemma 2*   The same applies to a set of the form $\{(x, b): a \le x \le a + d\}$.
□

Lemma 3   If $\mathbb{Q}$ is the union of a finite number of compact intervals in $\mathbb{R}^2$ and h is in $\mathbb{R}^2$ into $\mathbb{R}^2$ and is continuously differentiable at each point of $\mathbb{Q}$ and has an inverse continuous at each point of $\text{bdy}[h(\mathbb{Q})]$, then $h(\mathbb{Q})$ has content.

Proof $\text{bdy}[h(\mathbb{Q})] \subseteq h(\text{bdy } \mathbb{Q})$ by lemma 1, and so has zero content, by lemmas 2 and 2*.
□

Problem   16. If $\mathbb{Q}$ is a compact set with content and h is continuous on $\mathbb{Q}$, does $h(\mathbb{Q})$ necessarily have content?

Improper integrals

We have defined integrals only for bounded functions over bounded ranges of integration. Sometimes, however, we want to integrate over unbounded ranges, e.g. in potential theory and in statistics. Sometimes we would also like to integrate unbounded functions. For example, if $0 \le \beta < 1$, the length of the circular arc PQ in Fig. 5.4 is $\int^{\beta}_{-\beta} f$ where $f(x) = (1 - x^2)^{-1/2}$. We cannot, however, say that the length of the semicircle AB is $\int^{1}_{-1} f$, because the integrand is not bounded on $(-1; 1)$.

The problems of defining an integral over an unbounded range and of defining an integral of an unbounded function are closely similar. If the function is in $\mathbb{R}$ into $\mathbb{R}$ and the integral is interpreted as the 'area under the graph', then in each case we want to define the area of a certain unbounded region.

We have been talking about functions in $\mathbb{R}$ into $\mathbb{R}$ for simplicity, but similar remarks apply to functions in $\mathbb{R}^n$ into

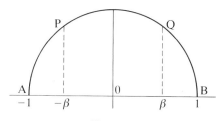

Fig. 5.4

$\mathbb{R}$ in general. We shall treat the case $n = 2$ in detail; our results are easily generalized.

Notation We denote the sequence $\mathbb{A}_1, \mathbb{A}_2, \ldots$ by $\{\mathbb{A}_v\}$.

Definition If $\mathbb{G}$ is a subset of $\mathbb{R}^2$, a sequence $\{\mathbb{A}_v\}$ **exhausts** $\mathbb{G}$ if each $\mathbb{A}_v$ is a compact subset of $\mathbb{G}$ with content, if $\mathbb{A}_v \subseteq \mathbb{A}_{v+1}$ for each v, and if each compact subset of $\mathbb{G}$ is contained in some $\mathbb{A}_v$. (Clearly $\mathbb{G}$ is then the union of the $\mathbb{A}_v$.) $\qquad\square$

Theorem If $\mathbb{G}$ is an open subset of $\mathbb{R}^2$, there is a sequence that exhausts $\mathbb{G}$, each set of which is the union of a finite number of non-overlapping squares (i.e. compact square intervals).

Proof Let $\{\mathbb{I}_v^1\}$ be a sequence of non-overlapping squares of side 1 that cover $\mathbb{R}^2$. If $\mathbb{G} = \mathbb{R}^2$, this is the sequence required; if not, number those squares of the sequence that are contained in $\mathbb{G}$ to form the sequence $\{\mathbb{J}_v^1\}$. Number the remainder as $\{\mathbb{K}_v^1\}$.

Quarter each of the squares $\mathbb{K}_v^1$ into squares of side $\frac{1}{2}$. Number those that are contained in $\mathbb{G}$ as $\{\mathbb{J}_v^2\}$ and the remainder as $\{\mathbb{K}_v^2\}$. Quarter each of the squares $\mathbb{K}_v^2$, and so on. Eventually we have, for each positive integer μ, sequences $\{\mathbb{J}_v^\mu\}$ and $\{\mathbb{K}_v^\mu\}$ of squares of side $2^{1-\mu}$. Now arrange the sequences $\{\mathbb{J}_v^1\} \{\mathbb{J}_v^2\} \ldots$ in a single sequence $\{\mathbb{J}_v\}$ and put $\mathbb{A}_\kappa = \bigcup_{v=1}^\kappa \mathbb{J}_v$. Then each $\mathbb{A}_v$ is the union of v non-overlapping squares.

Let $\mathbb{K}$ be any non-null compact subset of $\mathbb{G}$ and ξ be any point of $\mathbb{K}$. Because $\mathbb{G}$ is open, there is a neighbourhood $\mathbb{U}$ with centre ξ contained in $\mathbb{G}$. There is a μ such that every square of side $2^{1-\mu}$ that contains ξ will be contained in $\mathbb{U}$. Then ξ cannot belong to $\mathbb{K}_v^\mu$ for any v. If ξ does not belong to some $\mathbb{J}_v^1$ it belongs to some $\mathbb{K}_v^1$ and therefore to some $\mathbb{J}_v^2$ or some $\mathbb{K}_v^2$; similarly if it does not belong to some $\mathbb{J}_v^2$ it belongs to some $\mathbb{J}_v^3$ or some $\mathbb{K}_v^3$, and so on. Therefore if it does not belong to some $\mathbb{J}_v^\lambda$ with $\lambda < \mu$, it belongs to some $\mathbb{J}_v^\mu$ or some $\mathbb{K}_v^\mu$ and therefore (because it does not belong to any $\mathbb{K}_v^\mu$) to some $\mathbb{J}_v^\mu$.

The complement of $\mathbb{G}$ is closed and non-null, and so, by the corollary on p. 34, there is an α in $\mathbb{K}$ and a β in the complement of $\mathbb{G}$ such that

$$|\alpha - \beta| \le |\zeta - \eta| \quad \text{whenever} \quad \zeta \in \mathbb{K} \text{ and } \eta \notin \mathbb{G}.$$

We have seen that ξ belongs to some $\mathbb{J}_\nu^\lambda$. This will be either some $\mathbb{I}_\nu^1$ or a quarter of some $\mathbb{K}_\nu^{\lambda-1}$. $\mathbb{K}_\nu^{\lambda-1}$ contains a point not in $\mathbb{G}$, so its diagonal is greater than $|\alpha - \beta|$. Therefore its side is greater than $|\alpha - \beta|/\sqrt{2}$.

Thus ξ lies in some $\mathbb{J}_\nu^\lambda$ whose side is at least $\min\{1, |\alpha - \beta|/2\sqrt{2}\}$. Thus if $2^{1-\lambda} \leq \min\{1, |\alpha - \beta|/2\sqrt{2}\}$, the sequences

$$\{\mathbb{J}_\nu^1\}, \{\mathbb{J}_\nu^2\}, \ldots, \{\mathbb{J}_\nu^\lambda\} \tag{13}$$

between them cover $\mathbb{K}$. Let $\mathbb{Q}_1$ be a square containing $\mathbb{K}$ and let $\mathbb{Q}_2$ have the same centre and sides greater by 2. Remove all the squares in (13) that lie outside $\mathbb{Q}_1$. None of these contain a point of $\mathbb{K}$, so the remainder cover $\mathbb{K}$; they are contained in $\mathbb{Q}_2$, because their sides are at most 1. They are non-overlapping and each has content at least $(2^{1-\lambda})^2$; therefore there are only a finite number of them. Then there will be an $\mathbb{A}_\kappa$ that contains their union and therefore contains $\mathbb{K}$. Therefore $\{\mathbb{A}_\nu\}$ exhausts $\mathbb{G}$. □

Definitions

If f is a function with values in $\mathbb{R}$, then

$$f^+ \text{ is } \tfrac{1}{2}(|f| + f)$$

and

$$f^- \text{ is } \tfrac{1}{2}(|f| - f).$$

Clearly the values of f^+ and f^- are all non-negative. In fact, $f^+(x) = f(x)$ if $f(x) \geq 0$, while $f^+(x) = 0$ if $f(x) \leq 0$, and vice versa for f^-. □

Theorem If f is integrable on $\mathbb{G}$ so are f^+ and f^-, and $\int_\mathbb{G} f = \int_\mathbb{G} f^+ - \int_\mathbb{G} f^-$.

Note It is obvious that f^+ is integrable, because it varies less than f, and so the gap between its upper and lower sums is smaller. The same applies to f^-, and the final result follows because $f^+ - f^- = f$. However, some details in the proof that f^+ is integrable may not be obvious, so this proof is given in full.

Proof Let $\mathbb{I}$ be a compact interval containing $\mathbb{G}$. Then by definition $\int_\mathbb{G} f = \int_\mathbb{I} g$ where $g = f_\mathbb{G}|_\mathbb{I}$. Thus if ϵ is any positive number, there is a partition $\mathscr{P}$ of $\mathbb{I}$ such that

$$\sum_{\mathbb{S} \in \mathscr{P}} (\sup_\mathbb{S} g - \inf_\mathbb{S} g) \cdot c(\mathbb{S}) < \epsilon. \tag{14}$$

If $g(x) \leq 0$ for every x in $\mathbb{S}$, then $g^+(x) = 0$ for every x in $\mathbb{S}$ and so

$$\sup_{\mathbb{S}} g^+ - \inf_{\mathbb{S}} g^+ = 0 \leq \sup_{\mathbb{S}} g - \inf_{\mathbb{S}} g.$$

If, however, $g(x) > 0$ for at least one x in $\mathbb{S}$, then $\sup_{\mathbb{S}} g^+ = \sup_{\mathbb{S}} g$. Clearly $\inf_{\mathbb{S}} g^+ \geq \inf_{\mathbb{S}} g$ and so again

$$\sup_{\mathbb{S}} g^+ - \inf_{\mathbb{S}} g^+ \leq \sup_{\mathbb{S}} g - \inf_{\mathbb{S}} g.$$

Then, by (14),

$$\sum_{\mathbb{S} \in \mathscr{P}} (\sup_{\mathbb{S}} g^+ - \inf_{\mathbb{S}} g^+) \cdot c(\mathbb{S}) < \epsilon.$$

This proves that g^+ is integrable. But

$$g^+ = f_{\mathbb{G}}^+|_{\mathbb{I}}$$

because their values at x are both $f(x)$ if $x \in \mathbb{G}$ and $f(x) \geq 0$, and both zero otherwise. Hence $f_{\mathbb{G}}^+|_{\mathbb{I}}$ is integrable, which means that f^+ is integrable on $\mathbb{G}$. $\qquad \square$

Theorem   Let $\mathbb{G}$ be a bounded set exhausted by $\{A_\nu\}$ and f be integrable on $\mathbb{G}$. Then

$$\int_{\mathbb{G}} f = \lim \left\{ \int_{A_\nu} f \right\}.$$

Proof   Case (i): The values of f are non-negative. By the theorem on p. 122, $\int_{A_\nu} f$ exists. Then $\{\int_{A_\nu} f\}$ is a non-decreasing sequence bounded above by $\int_{\mathbb{G}} f$. Therefore its limit exists; call it λ. Clearly $\lambda \leq \int_{\mathbb{G}} f$, so we need only prove that $\lambda \geq \int_{\mathbb{G}} f$. Let $\mathbb{I}$ be any compact interval containing $\mathbb{G}$, and $f_1 = f_{\mathbb{G}}|_{\mathbb{I}}$. There is a partition $\mathscr{P}$ of $\mathbb{I}$ such that

$$\sum_{\mathbb{S} \in \mathscr{P}} \inf_{\mathbb{S}} f_1 \cdot c(\mathbb{S}) > \int f_1 - \epsilon.$$

Let $\mathbb{K}$ be the union of those cells of $\mathscr{P}$ that are contained in $\mathbb{G}$. Then

$$\sum_{\mathbb{S} \in \mathscr{P}} \inf_{\mathbb{S}} f_1 \cdot c(\mathbb{S}) = \sum_{\mathbb{S} \subseteq \mathbb{K}} \inf_{\mathbb{S}} f_1 \cdot c(\mathbb{S}).$$

$\mathbb{K}$ is compact, and so there is a κ such that $\mathbb{A}_\kappa \supseteq \mathbb{K}$. Then

$$\lambda \geq \int_{\mathbb{A}_\kappa} f \geq \int_{\mathbb{K}} f$$

$$= \sum_{\mathbb{S} \subseteq \mathbb{K}} \int_{\mathbb{S}} f$$

$$\geq \sum_{\mathbb{S} \subseteq \mathbb{K}} \inf_{\mathbb{S}} f \cdot c(\mathbb{S})$$

$$\geq \sum_{\mathbb{S} \subseteq \mathbb{K}} \inf_{\mathbb{S}} f_1 \cdot c(\mathbb{S})$$

$$= \sum_{\mathbb{S} \in \mathscr{P}} \inf_{\mathbb{S}} f_1 \cdot c(\mathbb{S})$$

$$> \int f_1 - \epsilon$$

$$= \int_{\mathbb{G}} f - \epsilon.$$

This is true for every positive ϵ. Therefore

$$\lambda \geq \int_{\mathbb{G}} f.$$

Case (ii). The general case.

$$\int_{\mathbb{G}} f = \int_{\mathbb{G}} f^+ - \int_{\mathbb{G}} f^-$$

$$= \lim \left\{ \int_{\mathbb{A}_\nu} f^+ \right\} - \lim \left\{ \int_{\mathbb{A}_\nu} f^- \right\} \qquad \text{by case (i)}$$

$$= \lim \left\{ \int_{\mathbb{A}_\nu} (f^+ - f^-) \right\}$$

$$= \lim \left\{ \int_{\mathbb{A}_\nu} f \right\}. \qquad\qquad \square$$

Corollary   If $\mathbb{G}$ is a set with content exhausted by $\{\mathbb{A}_\nu\}$, then $c(\mathbb{G}) = \lim\{c(\mathbb{A}_\nu)\}$.

Proof   Let f be the constant function with domain $\mathbb{G}$ and value 1: apply the theorem.

Lemma If $\mathbb{G}$ is a subset of $\mathbb{R}^2$ exhausted by $\{\mathbb{A}_v\}$ and exhausted by $\{\mathbb{B}_v\}$, then either

$$\lim\{c(\mathbb{A}_v)\} = \lim\{c(\mathbb{B}_v)\}$$

or

$\{c(\mathbb{A}_v)\}$ and $\{c(\mathbb{B}_v)\}$ are both divergent.

Proof Each $\mathbb{A}_v$ is contained in some $\mathbb{B}_v$ and vice versa. (Note. In contrast with the corollary above, here the set $\mathbb{G}$ need not have content.)

Definition If $\mathbb{G}$ is a subset of $\mathbb{R}^2$ exhausted by $\{\mathbb{A}_v\}$, then the **content** $c(\mathbb{G})$ of $\mathbb{G}$ is
$$\lim\{c(\mathbb{A}_v)\}. \qquad \square$$

Note The lemma shows that the definition is unambiguous. If $\{c(\mathbb{A}_v)\}$ is divergent we say that $c(\mathbb{G})$ is "infinite". If $\mathbb{G}$ is open it has a content (possibly infinite). The main motive for this definition is to give a content to unbounded open sets. It may (and in fact does) give a content to some bounded sets whose content is hitherto undefined. It is compatible with the existing definition, as the corollary above shows.

Lemma If $\{\mathbb{A}_v\}$ and $\{\mathbb{B}_v\}$ each exhausts $\mathbb{G}$, if f is integrable on each $\mathbb{A}_v$ and on each $\mathbb{B}_v$, and if the values of f are all non-negative, then

$$\lim\left\{\int_{\mathbb{A}_v} f\right\} = \lim\left\{\int_{\mathbb{B}_v} f\right\}$$

or both $\{\int_{\mathbb{A}_v} f\}$ and $\{\int_{\mathbb{B}_v} f\}$ diverge.

Proof Each $\mathbb{A}_v$ is contained in some $\mathbb{B}_\mu$ and then $\int_{\mathbb{A}_v} f \leq \int_{\mathbb{B}_\mu} f$, and vice versa.

Definition (Part 1)

If $\mathbb{G}$ is a subset of $\mathbb{R}^2$ exhausted by $\{\mathbb{A}_v\}$, if f is integrable on every compact subset of $\mathbb{G}$ that has content, and if the values of f are all non-negative, then $\int_{\mathbb{G}} f$ is defined to be $\lim\{\int_{\mathbb{A}_v} f\}$. $\qquad \square$

Note The lemma shows that the definition is unambiguous. The main motive for this definition is to define integrals over unbounded open sets. It is compatible with the previous definition for bounded sets as the theorem on p. 128 shows.

Definition (Part 2)

If $\int_\mathbb{G} f^+$ and $\int_\mathbb{G} f^-$ both exist and are not both infinite, then we define $\int_\mathbb{G} f$ to be $\int_\mathbb{G} f^+ - \int_\mathbb{G} f^-$.   □

Note

The theorem on p. 127 shows that if $\int_\mathbb{G} f$ exists under the old definition, then it exists under the new definition and the two integrals are the same. If $\int_\mathbb{G} f$ exists under the new definition but not under the old definition, it is called an **improper** integral. The word 'integrable' now has a wider meaning than before. If $\int_\mathbb{G} f$ exists under the new definition but not under the old one, we say that f is **improperly** integrable on $\mathbb{G}$; if it exists under the old definition, f is **properly** integrable on $\mathbb{G}$.

Theorem

If $\int_\mathbb{G} f$ exists and $\{\mathbb{A}_v\}$ exhausts $\mathbb{G}$, then $\int_\mathbb{G} f = \lim\{\int_{\mathbb{A}_v} f\}$.

Proof

$$\int_\mathbb{G} f = \int_\mathbb{G} f^+ - \int_\mathbb{G} f^- \qquad \text{by definition}$$

$$= \lim\left\{\int_{\mathbb{A}_v} f^+\right\} - \lim\left\{\int_{\mathbb{A}_v} f^-\right\} \qquad \text{by definition}$$

$$= \lim\left\{\int_{\mathbb{A}_v} (f^+ - f^-)\right\}$$

$$= \lim\left\{\int_{\mathbb{A}_v} f\right\}. \qquad\qquad\qquad \square$$

Examples

13. Let $f(x, y) = x^{-1/2}$ whenever $x \neq 0$, and let $\mathbb{G} = (0; 1) \times (0; 1)$. Then f is not bounded on $\mathbb{G}$. Set $\mathbb{A}_v = \mathbb{I}_v \times \mathbb{I}_v$ where $\mathbb{I}_v = [1/(v + 2); 1 - 1/(v + 2)]$ for each positive integer v. (Thus the $\mathbb{A}_v$ form an expanding sequence of squares with centres at the centre of $\mathbb{G}$). If $\mathbb{K}$ is any compact subset of $\mathbb{G}$, then, because the complement of $\mathbb{G}$ is closed, there is, by the corollary on p. 34, a number d such that

$$|\xi - \eta| \geq d$$

whenever ξ lies in $\mathbb{K}$ and η lies in the complement of $\mathbb{G}$. Then any point ξ in $\mathbb{K}$ must be at least d from the boundary of $\mathbb{G}$, and therefore lies in $\mathbb{A}_v$ if v is any integer greater than $1/d$. Therefore $\{\mathbb{A}_v\}$ exhausts $\mathbb{G}$.

By corollary (3) to the iterated-integral theorem (p. 113)

$$\int_{\mathbb{A}_v} f = \int_{\mathbb{I}_v} u \qquad \text{where} \qquad u(x) = \int_{\mathbb{I}_v} f(x, \cdot).$$

Now $f(x, \cdot)$ is constant with value $x^{-1/2}$, and so

$$u(x) = x^{-1/2}\left(1 - \frac{2}{v+2}\right).$$

Then

$$\int_{\mathbb{A}_v} f = 2\left\{\left(1 - \frac{1}{v+2}\right)^{1/2} - \frac{1}{(v+2)^{1/2}}\right\} \cdot \left\{1 - \frac{2}{v+2}\right\}$$

and the limit of this is 2. Hence $\int_{\mathbb{G}} f = 2$.

14. Let $f(x, y) = 1/(1 + x^2)(1 + y^2)$, and $\mathbb{G} = \{(x, y) : x \geq 0$ and $y \geq 0\}$. Set $\mathbb{A}_v = [0; v] \times [0; v]$. Any compact subset of $\mathbb{G}$ is bounded and therefore contained in some $\mathbb{A}_v$. Therefore $\{\mathbb{A}_v\}$ exhausts $\mathbb{G}$.

$$\int_{\mathbb{A}_v} f = \int_{[0;v]} u \qquad \text{where} \qquad u(x) = \int_{[0;v]} f(x, \cdot)$$
$$= [(1 + x^2)^{-1} \arctan y]_{y=0}^{y=v}$$
$$= (1 + x^2)^{-1} \arctan v.$$

Then

$$\int_{\mathbb{A}_v} f = (\arctan v)[\arctan x]_{x=0}^{x=v}$$
$$= (\arctan v)^2.$$

Then

$$\int_{\mathbb{G}} f = (\pi/2)^2.$$

Exercises 23. Evaluate $\int_{\mathbb{G}} f$ where $\mathbb{G} = \{(x, y) : 0 < x < 1$ and $0 < y < x\}$ and $f(x, y) = y/x^2$ whenever $x \neq 0$.

24. Evaluate $\int_{\mathbb{G}} f$ where $\mathbb{G} = \{(x, y) : 1 < x < y\}$ and $f(x, y) = 1/xy^2$ whenever $xy \neq 0$.

Problems 17. As Exercises 23 and 24, but in each case take $f(x, y) = x^m y^n$ whenever $xy \neq 0$.

18. Let $\{\mathbb{I}_v\}$ be the squares into which $\mathbb{R}^2$ is divided by the lines $x = m$ and $y = m$ for all integers m, arranged in a sequence. Let $f(\xi) = (-1)^v / v$ if ξ is interior to $\mathbb{I}_v$, and $f(\xi) = 0$ if ξ is on the edge of a square. Let $\mathbb{A}_v = \bigcup_{i=1}^{v} \mathbb{I}_i$. Show that $\{\mathbb{A}_v\}$ exhausts $\mathbb{R}^2$, that $\lim\{\int_{\mathbb{A}_v} f\}$ exists, and that $\int_{\mathbb{R}^2} f$ does not exist.

19. $\mathbb{G}$ is an open subset of $\mathbb{R}^2$. f and g are integrable on $\mathbb{G}$. Which of

the following are necessarily integrable on $\mathbb{G}$:

$$f + g, \quad f - g, \quad fg, \quad |f|, \quad |fg|, \quad (f^2 + g^2)^{1/2}?$$

The content of plane sets and transformations of the plane

We now investigate the effect of mappings on contents of sets; that is to say, if $\mathbb{S}$ is a subset of $\mathbb{R}^2$ and h a mapping, we ask what can be said about the content of $h(\mathbb{S})$.

Very informally, let $\mathbb{I}$ be a compact interval in $\mathbb{R}^2$ and h be a function in $\mathbb{R}^2$ into $\mathbb{R}^2$ whose domain includes $\mathbb{I}$. If h is not too irregular and if $\mathbb{I}$ is small, $h(\mathbb{I})$ will be almost a parallelogram. Let us interpolate a relevant exercise here.

Exercise   25. Let two opposite corners of $\mathbb{I}$ be at $(1, 1)$ and $(1.1, 1.1)$, and let $h(x, y) = (xy, x^2 - y^2)$. What is $h(\mathbb{I})$?

To continue, if two opposite corners of $\mathbb{I}$ are at α and $\alpha + \delta$, then the corners of $h(\mathbb{I})$ are at

$$h(\alpha) \tag{15}$$

$$h(\alpha_1 + \delta_1, \alpha_2) \approx h(\alpha) + (\delta_1 D_1 h_1(\alpha), \delta_1 D_1 h_2(\alpha)) \tag{16}$$

$$h(\alpha_1, \alpha_2 + \delta_2) \approx h(\alpha) + (\delta_2 D_2 h_1(\alpha), \delta_2 D_2 h_2(\alpha)) \tag{17}$$

$$h(\alpha + \delta) \tag{18}$$

and so its area is approximately that of a parallelogram, three of whose corners are $h(\alpha)$ and the right-hand sides of (16) and (17). The area of this parallelogram, by the computation used in Exercise 21, is

$$|\delta_1 \delta_2 \det \mathrm{D}h(\alpha)|.$$

This determinant plays an important role in analysis, and is named after Carl Jacobi.

Definition   If h is a function in $\mathbb{R}^2$ into $\mathbb{R}^2$, its **Jacobian** Jh is the function defined by

$$Jh(\xi) = |\det \mathrm{D}h(\xi)|$$

for every ξ for which $\mathrm{D}h(\xi)$ exists.   □

Thus the Jacobian is a kind of magnification factor. Our calculations have been very rough. We might hope that as δ

approaches $(0, 0)$ they might approach exactness, so that the Jacobian would be an exact magnification factor at a point. Then the integral of the Jacobian over a region should be the area of the image of the region:

$$c(h(\mathbb{D})) = \int_{\mathbb{D}} Jh \qquad (19)$$

under suitable conditions. This is in fact true but its complete proof is long and is bound up with the change-of-variable theorem for integrals, of which it is a special case. The left-hand side of (19) is

$$\int_{h(\mathbb{D})} \mathbf{1}$$

and this is the special case in which $f = \mathbf{1}$ of the integral

$$\int_{h(\mathbb{D})} f$$

Let us, then, investigate this integral, still arguing very informally. We divide $\mathbb{D}$ into small cells and take a typical point in each (as shown in Fig. 5.5(a)). Figure 5.5(b) represents the image under h of Fig. 5.5(a) and shows that $\int_{h(\mathbb{D})} f$ is approximately

$$\sum f(h(\alpha)) \cdot c(h(\mathbb{S})). \qquad (20)$$

However, by (19), $c(h(\mathbb{S}))$ is approximately $Jh(\alpha) \cdot c(\mathbb{S})$,

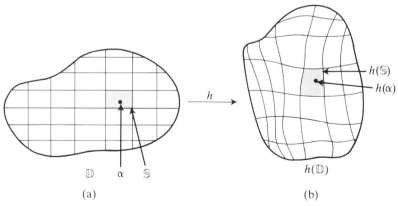

(a)   (b)

Fig. 5.5

and so (20) is approximately equal to

$$\sum f(h(\alpha)) \cdot Jh(\alpha) \cdot c(\mathbb{S}) \approx \int_{\mathbb{D}} f \circ h \cdot Jh.$$

This gives the change-of-variable formula $\int_{h(\mathbb{D})} f = \int_{\mathbb{D}} f \circ h \cdot Jh$. All this is treated formally in the rest of this chapter.

Exercises   26. Show that the Jacobian of a linear function is constant. Is the converse true?

27. If h is linear, if the value of Jh is λ, and if $\mathbb{I}$ is a compact interval, does it follow that $c(h(\mathbb{I})) = \lambda \cdot c(\mathbb{I})$?

28. Let cartesian coordinates be set up in a plane, let P be the point with coordinates (x, y), and let OP be rotated about O through an angle θ. Then P will be moved to the point with coordinates

$$(x \cos \theta - y \sin \theta, x \sin \theta + y \cos \theta).$$

Thus if we let

$$f(x, y) = (x \cos \theta - y \sin \theta, x \sin \theta + y \cos \theta),$$

we can say that the rotation moves a point with coordinates α to the position where the coordinates are $f(\alpha)$. Given a geometrical figure $\mathscr{F}$ in the plane, let $\mathbb{D}$ be the set of all coordinates of its points. Then $f(\mathbb{D})$ would be the set of all coordinates of points in the figure $\mathscr{F}^{\theta}$ obtained by rotating $\mathscr{F}$ about O through the angle θ. The figures $\mathscr{F}$ and $\mathscr{F}^{\theta}$ have the same area; thus if content is to act (as we have suggested) as an analogue of area for $\mathbb{R}^2$, we should expect $\mathbb{D}$ and $f(\mathbb{D})$ to have the same content. That is precisely the exercise: prove that if f is as defined above and $\mathbb{D} \subseteq \mathbb{R}^2$, then

$$c(\mathbb{D}) = c(f(\mathbb{D}))$$

if either exists.   □

Definition   If h is the linear function for which

$$h(x, y) = (ax + by, cx + dy),$$

then $\det h$ denotes

$$\det \begin{bmatrix} a & b \\ c & d \end{bmatrix}.$$

Thus Jh, which is constant, has value $|\det h|$.   □

Theorem   If $\eta = f(\xi)$, if f is continuously differentiable at ξ and h is continuously differentiable at η, and if g is the composite function $h \circ f$, then

$$Jg(\xi) = Jh(\eta) \cdot Jf(\xi).$$

Proof

$$
\begin{aligned}
Jg(\xi) &= |\det \mathrm{D}g(\xi)| \\
&= |\det[\mathrm{D}h(f(\xi)) \cdot \mathrm{D}f(\xi)]| \\
&= |\det \mathrm{D}h(\eta)| \cdot |\det \mathrm{D}f(\xi)| \\
&= Jh(\eta) \cdot Jf(\xi).
\end{aligned}
$$

$\square$

Corollary If h is continuously differentiable at η and its inverse h^- is continuously differentiable at $h(\eta)$, then

$$Jh(\eta) \cdot Jh^-(h(\eta)) = 1.$$

It follows that under these conditions

$$Jh(\eta) \neq 0.$$

$\square$

Definition If α, β, and γ belong to $\mathbb{R}^2$, then

$$\{\alpha + u\beta + v\gamma : 0 \le u \le 1, 0 \le v \le 1\}$$

is the **lozenge** spanned by β and γ at α. (In the usual diagram it is represented as a parallelogram.) $\square$

Notation If ξ belongs to $\mathbb{R}^2$, then ξ^* is the column matrix whose entries are the components of ξ in order. If ξ and η belong to $\mathbb{R}^2$, then $[\xi^*, \eta^*]$ is the matrix whose columns are ξ^* and η^*.

Theorem The content of a lozenge $\mathbb{P}$ spanned by β and γ is $|\det[\beta^*, \gamma^*]|$.

Proof Let $\mathbb{I}$ be a compact interval containing $\mathbb{P}$, and f be the function with domain $\mathbb{I}$ with value 1 on $\mathbb{P}$ and zero off $\mathbb{P}$. We evaluate $\int f$ by the iterated-integral theorem. The details are straightforward but tedious. $\square$

Corollary 1

If $\mathbb{P}$ is a lozenge and h is a linear function on $\mathbb{R}^2$ into $\mathbb{R}^2$, then $c(h(\mathbb{P})) = |\det h| \cdot c(\mathbb{P})$.

Proof $h(\alpha + u\beta + v\gamma) = h(\alpha) + u \cdot h(\beta) + v \cdot h(\gamma)$, and so $h(\mathbb{P})$ is the lozenge spanned by $h(\beta)$ and $h(\gamma)$ at $h(\alpha)$. Then

$$
\begin{aligned}
c(h(\mathbb{P})) &= |\det[h(\beta)^*, h(\gamma)^*]| \\
&= |\det[H \cdot \beta^*, H \cdot \gamma^*]| \quad \text{where } H \text{ is the matrix of } h \\
&= |\det H \cdot [\beta^*, \gamma^*]| \\
&= |\det H| \cdot |\det[\beta^*, \gamma^*]| \\
&= |\det h| \cdot c(\mathbb{P}). \quad \Box
\end{aligned}
$$

Corollary 2

If $\mathbb{P}$ is the union of a finite number of non-overlapping lozenges and h is linear, then $c(h(\mathbb{P})) = |\det h| \cdot c(\mathbb{P})$. $\Box$

Theorem   If $\mathbb{G}$ is open and h is a linear function on $\mathbb{R}^2$ onto $\mathbb{R}^2$ and has an inverse, then $c(h(\mathbb{G})) = |\det h| \cdot c(\mathbb{G})$.

Proof   Let $\{\mathbb{A}_\nu\}$ exhaust $\mathbb{G}$ and each $\mathbb{A}_\nu$ be the union of a finite number of squares. Each $h(\mathbb{A}_\nu)$ is a compact subset of $h(\mathbb{G})$ with content (by lemma 3 on p. 125) and is contained in $h(\mathbb{A}_{\nu+1})$. If $\mathbb{K}$ is any compact subset of $h(\mathbb{G})$ and h^- is the inverse of h, then $h^-\mathbb{K}$ is a compact subset of $\mathbb{G}$ and so is contained in some $\mathbb{A}_\nu$. Then $\mathbb{K} \subseteq h(\mathbb{A}_\nu)$. Therefore $\{h(\mathbb{A}_\nu)\}$ exhausts $h(\mathbb{G})$. Then

$$
\begin{aligned}
|\det h| \cdot c(\mathbb{G}) &= |\det h| \cdot \lim\{c(\mathbb{A}_\nu)\} \\
&= \lim\{c(h(\mathbb{A}_\nu))\} \quad \text{by corollary 2 above} \\
&= c(h(\mathbb{G})). \quad \Box
\end{aligned}
$$

Theorem   If h is a linear function on $\mathbb{R}^2$ onto $\mathbb{R}^2$ with an inverse, if $\mathbb{Q}$ is a compact interval in $\mathbb{R}^2$, if g is in $\mathbb{R}^2$ into $\mathbb{R}^2$ and is continuously differentiable and has a continuously differentiable inverse, and if $\text{dom } g \supseteq \mathbb{Q}$, then

$$
c(h \circ g(\mathbb{Q})) = |\det h| \cdot c(g(\mathbb{Q})).
$$

Proof   Let $\mathbb{P}$ be the interior of $\mathbb{Q}$. Then $g(\mathbb{P})$ is open (because the inverse g^- of g is continuous) and so

$$
c(h(g(\mathbb{P}))) = |\det h| \cdot c(g(\mathbb{P})).
$$

However,

$$
\begin{aligned}
g(\mathbb{Q}) &= g(\mathbb{P} \cup \text{bdy } \mathbb{Q}) \\
&= g(\mathbb{P}) \cup g(\text{bdy } \mathbb{Q})
\end{aligned}
$$

and, by lemmas 2 and 2* on pp. 124, 125, $g(\text{bdy } \mathbb{Q})$ has content zero. Therefore

$$c(g(\mathbb{Q})) = c(g(\mathbb{P})).$$

Similarly

$$c(h \circ g(\mathbb{Q})) = c(h \circ g(\mathbb{P})).$$

The 'Jacobian as a magnification factor' theorem

Let g be a continuously differentiable function in $\mathbb{R}^2$ into $\mathbb{R}^2$ with a continuously differentiable inverse. Let $\mathbb{K}$ be a compact subset of $\text{dom}\, g$ and let $\epsilon > 0$. Then there is a positive δ such that

$$\left| \frac{c(g(\mathbb{Q}))}{c(\mathbb{Q})} - Jg(\xi) \right| < \epsilon$$

whenever $\xi \in \mathbb{K}$ and $\mathbb{Q}$ is a square with centre ξ and side less than 2δ contained in $\mathbb{K}$. (Informally, $c(g(\mathbb{Q}))/c(\mathbb{Q}) \to Jg(\xi)$ as $\mathbb{Q}$ shrinks to ξ, and the convergence is uniform on $\mathbb{K}$.)

Proof   Let ϵ be any positive number. If $\xi \in \mathbb{K}$, let l_ξ be defined by

$$l_\xi(\tau)^* = Dg(\xi) \cdot \tau^*$$

(see p. 97). Then its inverse l_ξ^- satisfies

$$l_\xi^-(\tau)^* = Dg^-(g(\xi)) \cdot \tau^*.$$

The $D_j g_i^-$ are continuous on the compact set $g(\mathbb{K})$ and consequently are bounded on $g(\mathbb{K})$. Therefore, by the theorem on p. 98 there is a μ such that

$$l_\xi^-(\tau) \le \mu \cdot |\tau| \qquad \text{whenever } \xi \in \mathbb{K}. \tag{21}$$

Jg is continuous on the compact set $\mathbb{K}$ and therefore bounded on $\mathbb{K}$: let λ be an upper bound. There is a γ between 0 and 1 for which

$$1 - \epsilon/\lambda \le (1 - \gamma)^2 \le (1 + \gamma)^2 \le 1 + \epsilon/\lambda. \tag{22}$$

By the lemma on p. 98 there is a number δ such that

$$|g(\xi + \eta) - g(\xi) - l_\xi(\eta)| < \tfrac{1}{2} |\eta| \cdot \frac{\gamma}{\mu}$$

whenever $|\eta| < 2\delta$, $\xi \in \mathbb{K}$, and $\xi + \eta \in \mathbb{K}$. Then

$$
\begin{aligned}
|l_\xi^-(g(\xi + \eta)) &- l_\xi^-(g(\xi)) - \eta| \\
&= |l_\xi^-[g(\xi + \eta) - g(\xi) - l_\xi(\eta)]| \\
&\le \mu \cdot |g(\xi + \eta) - g(\xi) - l_\xi(\eta)| \qquad \text{by (21)} \\
&< \tfrac{1}{2} |\eta| \cdot \gamma.
\end{aligned}
\tag{23}
$$

Now let $0 < \sigma < \delta$, let $\mathbb{Q}$ be the square with centre ξ and side 2σ, and let $\mathbb{Q}_1$, $\mathbb{Q}_2$, and $\mathbb{Q}_3$ be the squares with centre $l_\xi^\sim(g(\xi))$ and sides $2\sigma(1 + \gamma)$, $2\sigma(1 - \gamma)$, and 2σ respectively. Then if $\xi + \eta$ is on the edge of $\mathbb{Q}$,

$$|\eta| < 2\sigma < 2\delta$$

and $l_\xi^\sim(g(\xi)) + \eta$ is on the edge of $\mathbb{Q}_3$. By (23), if $\mathbb{Q} \subseteq \mathbb{K}$, then

$$l_\xi^\sim(g(\xi + \eta))$$

is within $\frac{1}{2}|\eta| \cdot \gamma$, and therefore within $\sigma\gamma$, of

$$l_\xi^\sim(g(\xi)) + \eta$$

and so lies inside $\mathbb{Q}_1$ and outside $\mathbb{Q}_2$. Therefore

$$\mathbb{Q}_2 \subseteq l_\xi^\sim(g(\mathbb{Q})) \subseteq \mathbb{Q}_1.$$

Then

$$4\sigma^2(1 - \gamma)^2 \le c(l_\xi^\sim(g(\mathbb{Q}))) \le 4\sigma^2(1 + \gamma)^2.$$

By the first theorem on page 137

$$c(l_\xi^\sim(g(\mathbb{Q}))) = \frac{c(g(\mathbb{Q}))}{Jg(\xi)}$$

because $l_\xi^\sim$ is linear and its determinant is $1/Jg(\xi)$. Then, by (22),

$$1 - \frac{\epsilon}{\lambda} < \frac{c(g(\mathbb{Q}))}{c(\mathbb{Q}) \cdot Jg(\xi)} < 1 + \frac{\epsilon}{\lambda}$$

because $c(\mathbb{Q}) = 4\sigma^2$. Then

$$\left| Jg(\xi) - \frac{c(g(\mathbb{Q}))}{c(\mathbb{Q})} \right| < Jg(\xi) \cdot \frac{\epsilon}{\lambda}$$

$$\le \epsilon. \qquad \Box$$

Lemma If g is in $\mathbb{R}^2$ into $\mathbb{R}^2$ and is continuously differentiable, and has a continuously differentiable inverse, and if $\mathbb{Q}$ is a square contained in dom g, then

$$c(g(\mathbb{Q})) = \int_{\mathbb{Q}} Jg.$$

Proof Let ϵ be any positive number. Then there is a δ_1 such that

$$\left| Jg(\xi) - \frac{c(g(\mathbb{K}))}{c(\mathbb{K})} \right| < \frac{\frac{1}{2}\epsilon}{c(\mathbb{Q})}$$

whenever $\mathbb{K}$ is a square with centre ξ and side less than δ_1 contained in $\mathbb{Q}$. There is also a δ_2 such that the upper and lower sums for $\boldsymbol{J}g$ are within $\frac{1}{2}\epsilon$ of $\int_\mathbb{Q} \boldsymbol{J}g$ whenever $\mathcal{P}$ is a partition of $\mathbb{Q}$ with mesh less than δ_2.

Let $\mathcal{P}$ be a partition of $\mathbb{Q}$ into squares, of mesh less than $\min(\delta_1, \delta_2)$, and let the cells be $\mathbb{S}_1, \mathbb{S}_2, \ldots$, and the centre of $\mathbb{S}_v$ be ξ_v. Then

$$\sum_v \boldsymbol{J}g(\xi_v) \cdot c(\mathbb{S}_v)$$

is between the upper and lower sums for $\boldsymbol{J}g$ and is therefore within $\frac{1}{2}\epsilon$ of $\int_\mathbb{Q} \boldsymbol{J}g$.

Also,

$$|c(g(\mathbb{S}_v))/c(\mathbb{S}_v) - \boldsymbol{J}g(\xi_v)| < \tfrac{1}{2}\epsilon/c(\mathbb{Q})$$

and so

$$|c(g(\mathbb{S}_v)) - \boldsymbol{J}g(\xi_v) \cdot c(\mathbb{S}_v)| < \tfrac{1}{2}\epsilon \cdot \frac{c(\mathbb{S}_v)}{c(\mathbb{Q})}$$

and so

$$\left| \sum_v c(g(\mathbb{S}_v)) - \sum_v \boldsymbol{J}g(\xi_v) \cdot c(\mathbb{S}_v) \right| < \tfrac{1}{2}\epsilon.$$

Therefore $\sum_v c(g(\mathbb{S}_v))$, which equals $c(g(\mathbb{Q}))$, lies within ϵ of $\int_\mathbb{Q} \boldsymbol{J}g$. But ϵ was any positive number. Therefore $c(g(\mathbb{Q})) = \int_\mathbb{Q} \boldsymbol{J}g$. $\qquad\square$

Theorem   Let g be in $\mathbb{R}^2$ into $\mathbb{R}^2$, be continuously differentiable, and have a continuously differentiable inverse. Let f be in $\mathbb{R}^2$ into $\mathbb{R}$ and $\operatorname{dom} f \supseteq \operatorname{rng} g$. Then

$$\int_{\operatorname{rng} g} f = \int_{\operatorname{dom} g} f \circ g \cdot \boldsymbol{J}g$$

if either integral exists.

Proof   Case (i). Suppose that the values of f are all non-negative and $\int_{\operatorname{dom} g} f \circ g \cdot \boldsymbol{J}g$ exists. Let $\mathbb{Q}$ be a square contained in $\operatorname{dom} g$. Let ϵ be any positive number. Let $\mathcal{P}$ be a partition of $\mathbb{Q}$ into squares for which

$$\ast\!\!\sum_\mathcal{P} f \circ g \cdot \boldsymbol{J}g \leq \int_\mathbb{Q} f \circ g \cdot \boldsymbol{J}g + \epsilon. \tag{24}$$

(The second theorem on p. 108 shows that such a $\mathcal{P}$ exists.)

If $\mathbb{S}$ is a cell of $\mathcal{P}$ then, by the preceding lemma,

$$c(g(\mathbb{S})) = \int_\mathbb{S} \boldsymbol{J}g.$$

Now $\sup_{g(\mathbb{S})} f = \sup_{\mathbb{S}} f \circ g$; if we multiply by this and sum over all cells $\mathbb{S}$ of $\mathcal{P}$, we find that

$$\sum (\sup_{g(\mathbb{S})} f) \cdot c(g(\mathbb{S})) = \sum \int_{\mathbb{S}} \sup_{\mathbb{S}} f \circ g \cdot \boldsymbol{J}g$$

$$\leq \sum \int_{\mathbb{S}} \sup_{\mathbb{S}} (f \circ g \cdot \boldsymbol{J}g)$$

$$\leq \sum \sup_{\mathbb{S}} (f \circ g \cdot \boldsymbol{J}g) \cdot c(\mathbb{S})$$

$$= {}^* \sum_{\mathcal{P}} f \circ g \cdot \boldsymbol{J}g$$

$$\leq \int_{\mathbb{Q}} f \circ g \cdot \boldsymbol{J}g + \epsilon \qquad \text{by (24).}$$

But

$${}^* \int_{g(\mathbb{Q})} f = \sum {}^* \int_{g(\mathbb{S})} f \leq \sum (\sup_{g(\mathbb{S})} f) \cdot c(g(\mathbb{S})).$$

Therefore

$${}^* \int_{g(\mathbb{Q})} f \leq \int_{\mathbb{Q}} f \circ g \cdot \boldsymbol{J}g + \epsilon$$

for every positive ϵ. Therefore

$${}^* \int_{g(\mathbb{Q})} f \leq \int_{\mathbb{Q}} f \circ g \cdot \boldsymbol{J}g.$$

We prove similarly that

$$\int_{* g(\mathbb{Q})} f \geq \int_{\mathbb{Q}} f \circ g \cdot \boldsymbol{J}g.$$

Therefore f is integrable on $g(\mathbb{Q})$, and

$$\int_{g(\mathbb{Q})} f = \int_{\mathbb{Q}} f \circ g \cdot \boldsymbol{J}g. \qquad (25)$$

Because $\mathrm{dom}\, g$ is open there is a sequence $\{\mathbb{A}_v\}$ that exhausts $\mathrm{dom}\, g$ in which each $\mathbb{A}_v$ is the union of a finite number of non-overlapping squares. Then, by applying (25) to each of the squares $\mathbb{Q}$ that compose $\mathbb{A}_v$ and adding, we see that

$$\int_{g(\mathbb{A}_v)} f = \int_{\mathbb{A}_v} f \circ g \cdot \boldsymbol{J}g.$$

Then

$$\lim\left\{\int_{g(\mathbb{A}_\nu)} f\right\} = \int_{\text{dom } g} f \circ g \cdot \mathbf{J}g.$$

The $g(\mathbb{A}_\nu)$ are compact subsets of rng g with content, by lemma 3 on p. 125, and $g(\mathbb{A}_\nu) \subseteq g(\mathbb{A}_{\nu+1})$. If $\mathbb{K}$ is any compact subset of rng g, then $g^-\mathbb{K}$ is a compact subset of dom g and so is contained in some $\mathbb{A}_\mu$. Then $\mathbb{K} \subseteq g(\mathbb{A}_\mu)$. Therefore $\{g(\mathbb{A}_\nu)\}$ exhausts rng g, and so

$$\lim\left\{\int_{g(\mathbb{A}_\nu)} f\right\} = \int_{\text{rng } g} f.$$

Case (ii): $\int_{\text{dom } g} f \circ g \cdot \mathbf{J}g$ exists; the values of f are not necessarily non-negative. Then

$$(f \circ g \cdot \mathbf{J}g)^+ = f^+ \circ g \cdot \mathbf{J}g.$$

Therefore $\int_{\text{dom } g} f^+ \circ g \cdot \mathbf{J}g$ exists and so, by case (i), it equals $\int_{\text{rng } g} f^+$. A similar argument holds for f^-. Then

$$\int_{\text{rng } g} f = \int_{\text{rng } g} (f^+ - f^-)$$

$$= \int_{\text{dom } g} (f^+ \circ g \cdot \mathbf{J}g - f^- \circ g \cdot \mathbf{J}g)$$

$$= \int_{\text{dom } g} f \circ g \cdot \mathbf{J}g.$$

Case (iii): $\int_{\text{rng } g} f$ exists. Apply case (ii) to $f \circ g \cdot \mathbf{J}g$ in place of f and the inverse g^- of g in place of g. Then

$$\int_{\text{dom } g^-} (f \circ g \cdot \mathbf{J}g) \circ g^- \cdot \mathbf{J}g^- = \int_{\text{rng } g^-} f \circ g \cdot \mathbf{J}g,$$

i.e.

$$\int_{\text{rng } g} f = \int_{\text{dom } g} f \circ g \cdot \mathbf{J}g. \qquad \square$$

Example 15. In $\mathbb{R}$ the change-of-variable theorem is most often used to transform a difficult integrand into manageable form. It can also be used for this purpose in $\mathbb{R}^2$, but it is more often used to transform the range of integration into manageable form. For example, let us suppose that we want to evaluate

$$\int_{\mathbb{P}} f$$

where $f(x, y) = x^2$ and $\mathbb{P} = \{(x, y) : x^2 + y^2 \leq 1\}$. If we let

$$g(r, \theta) = (r \cos \theta, r \sin \theta)_{\mathbb{D}}$$

for every (r, θ) in $\mathbb{D}$, where

$$\mathbb{D} = (0; 1) \times (-\pi; \pi),$$

then g is continuously differentiable and has a continuously differentiable inverse, $\boldsymbol{J}g(r, \theta) = |r|$, and $\mathbb{P} \backslash g(\mathbb{D})$ has zero content. Therefore

$$\int_{\mathbb{P}} f = \int_{g(\mathbb{D})} f = \int_{\mathbb{D}} f \circ g \cdot \boldsymbol{J}g$$

$$= \int_{\mathbb{D}} h$$

where $h(r, \theta) = |r^3| \cos^2 \theta$. This is easily evaluated by iterated integration. Its value is $\pi/4$. (In traditional notation it is $\int_{r=0}^{r=1} \int_{\theta=-\pi}^{\theta=\pi} |r^3| \cos^2 \theta \, dr \, d\theta$.)

The reader will recognize the choice of g as the introduction of polar coordinates, suggested by the fact that diagrammatically $\mathbb{P}$ is circular.

Note A fairly common form of the theorem on p. 140 proves the formula only on condition that f and g are well behaved on some set containing $\mathbb{D}$ and its boundary. That form of the theorem could not be used for this example because g is not one-to-one on the set consisting of $\mathbb{D}$ and its boundary.

Exercises 29. $\mathbb{A} = \{(x, y) : 1 \leq x - y \leq 2$ and $2 \leq x + y \leq 4\}$. Evaluate $\int_{\mathbb{A}} f$ where $f(x, y) = ax^2 + by^2 + cx + dy$ for every x and y of $\mathbb{R}$.

30. $\mathbb{A} = \{(x, y) : x \geq 0, \ y \geq 0, \ 3 \leq x^2 + 2y^2 \leq 4$ and $1 \leq x^2 - 3y^2 \leq 2\}$. Evaluate $\int_{\mathbb{A}} f$ where $f(x, y) = xy$ for every x and y in $\mathbb{A}$.

31. $\mathbb{A} = \{(x, y) : 3 \leq x^2 + 2y^2 \leq 4$ and $1 \leq x^2 - 3y^2 \leq 2\}$. Evaluate $\int_{\mathbb{A}} f$ where $f(x, y) = xy$ for every x and y in $\mathbb{R}$.

32. $\mathbb{A} = \{(x, y) : x > 0, \ y < 0, \ 3 \leq x^2 + 2y^2 \leq 4$ and $1 \leq x^2 - 3y^2 \leq 2\}$. Evaluate $\int_{\mathbb{A}} f$ where $f(x, y) = xy$ for every x and y in $\mathbb{R}$.

33. $\mathbb{A} = \{(x, y) : x > 0, \ xy > 1, \ xy < 2, \ x^2 - y^2 > 1$ and $x^2 - y^2 < 2\}$. Evaluate $\int_{\mathbb{A}} f$ where $f(x, y) = x^{n+4} y^n - x^n y^{n+4}$ for every x and y of $\mathbb{R}$.

34. $\mathbb{A} = \{(x, y) : x > 0, \ y > 0$ and $x + y < 1\}$. Evaluate $\int_{\mathbb{A}} f$ where

$$f(x, y) = \cos \frac{x - y}{x + y}$$

for every (x, y) in $\mathbb{A}$.

It is no coincidence that the change-of-variable theorem is presented for improper integrals. It may easily happen that

$$\int_{\mathrm{rng}\,g} f$$

exists as a proper integral, with f and $\mathrm{rng}\,g$ both bounded, while

$$\int_{\mathrm{dom}\,g} f \circ g \cdot Jg$$

exists only as an improper integral, with $\mathrm{dom}\,g$ or the integrand unbounded. Thus if we try to evaluate an integral by using the change-of-variable theorem, we may find that we have replaced it by an improper integral. The reverse can also happen.

Examples are most easily given in one dimension. Here t denotes the identity function on $\mathbb{R}$.

Examples 16. f is constant with value 2 and g is the restriction of $\arccos t^{1/2}$ to $(0; 1)$. Then

$$\int_{\mathrm{rng}\,g} f = \int_0^{\pi/2} 2 \qquad\qquad \text{proper}$$

$$\int_{\mathrm{dom}\,g} f \circ g \cdot Jg = \int_0^1 t^{-1/2}(1-t)^{-1/2} \qquad\qquad \text{improper (unbounded integrand).}$$

17. f is constant with value 2 and g is the restriction of $\arctan$ to $\{x : x > 0\}$. Then

$$\int_{\mathrm{rng}\,g} f = \int_0^{\pi/2} 2 \qquad\qquad \text{proper}$$

$$\int_{\mathrm{dom}\,g} f \circ g \cdot Jg = \int_0^\infty 2/(1+t^2) \qquad\qquad \text{improper (unbounded range).}$$

18. $f = 2/(1+t^2)$ and g is the restriction of $\tan$ to $(0; \pi/2)$. Then

$$\int_{\mathrm{rng}\,g} f = \int_0^\infty 2/(1+t^2) \qquad\qquad \text{improper (unbounded range)}$$

$$\int_{\mathrm{dom}\,g} f \circ g \cdot Jg = \int_0^{\pi/2} 2 \qquad\qquad \text{proper.}$$

Exercises   35. Find f and g such that $\int_{\text{rng}\,g} f$ is an improper integral with an unbounded integrand while $\int_{\text{dom}\,g} f \circ g \cdot Jg$ is an improper integral with unbounded range of integration.

36. Can you find f and g such that $\int_{\text{dom}\,g} f \circ g \cdot Jg$ is a proper integral while both rng g and $f_{\text{rng}\,g}$ are unbounded?

Note   The notation used in the examples may be unfamiliar to some readers. For instance, the first integral in example 18 is traditionally written

$$\int_0^\infty 2(1 + t^2)^{-1} \, dt.$$

To 'transform' this into the other integral, we set

$$t = \tan u$$

(the tan being the function g). Then

$$dt = \sec^2 u \, du$$
$$= (1 + t^2) \, du.$$

Also, $u = 0$ when $t = 0$, and $u \to \pi/2$ from below as $t \to \infty$. Thus the integral becomes

$$\int_0^{\pi/2} 2(1 + t^2)^{-1}(1 + t^2) \, du = \int_0^{\pi/2} 2 \, du,$$

which is a traditional way of writing the second integral. The calculations here are exactly the same as those made above, but the traditional notation, with its change of 'variable of integration', shows how the theorem got its name.

Integration in many dimensions

Integration in $\mathbb{R}^n$ follows closely along the lines of integration in two dimensions. Most of the definitions and results can be taken over either word for word or with the replacement of 2 by n. For example, a partition of a compact interval $\mathbb{I}_1 \times \ldots \times \mathbb{I}_n$ in $\mathbb{R}^n$ will be of the form

$$\{\mathbb{S}_1 \times \ldots \times \mathbb{S}_n : \mathbb{S}_i \in \mathscr{P}_i \text{ for each } i\}$$

where $\mathscr{P}_i$ is a partition of $\mathbb{I}_i$. If ξ and η belong to the same cell of a partition of mesh b, then $|\xi - \eta| \leq b \cdot n^{1/2}$. Refinements are defined exactly as in $\mathbb{R}^2$, and so are upper and lower sums and integrals (including improper integrals).

Exercises   37. Let $A = [0; b_1] \times \ldots \times [0; b_n]$. Evaluate $\int_A f$ and $\int_A g$ where $f(x_1, \ldots, x_n) = x_1{}^m + \ldots + x_n{}^m$ and $g(x_1, \ldots, x_n) = x_1 x_2 \ldots x_n$.

38. Evaluate $\int_A f$ where

$A = \{(x, y, z) : |x + y| \le 1, |x - z| \le 1 \text{ and } |y - z| \le 1\}$ and $f(x, y, z) = ax^2 + by^2 + cz^2 + px + qy + rz$ for every x, y, and z of $\mathbb{R}$.

Problems   20. f is uniformly continuous on a subset A of $\mathbb{R}^n$ and $S = \{(\xi_1, \ldots, \xi_n, f(\xi)) : \xi \in A\}$. Prove that the content of S (in $\mathbb{R}^{n+1}$) is zero.

21. Investigate the existence of $\int_A f$ where $A = \{\xi : |\xi| > 1\}$ and $f(\xi) = |\xi|^r (|\xi| - 1)^{-1/2}$.

6
Further developments

We have been studying intermediate analysis. What is higher analysis like?

Higher analysis uses new kinds of integral, more powerful than the Riemann integral described in this text. The best known of these is the Lebesgue integral, and it is more powerful in the following sense: if the Riemann integral of f exists, then so does the Lebesgue integral, and the two integrals are equal. However, there are functions whose Riemann integrals do not exist but whose Lebesgue integrals do exist; for example, the function whose domain is $[0; 1] \times [0; 1]$ and whose value at (x, y) is 1 if x and y are both rational, zero if not.

Another development is to work in abstract spaces: instead of dealing with functions in $\mathbb{R}^n$ into $\mathbb{R}^m$ we deal with functions in E into F where the elements of E and F are not necessarily ordered sets of numbers. In fact, they are not anything in particular; they are left unspecified. Instead of saying what *kinds* of things the elements of E and F are, we postulate the *properties* that they must have: for example, each pair of elements of E must have a 'sum' obeying such laws as the associative law. Many details of analysis in abstract spaces are the same as in the specific ordered-sets-of-numbers spaces, but some are not, because it is not generally possible to split elements of abstract spaces into components. Thus there is no such thing as the matrix of derivatives, and indeed the concept of derivative has to be modified. This is how the concept of the Fréchet derivative (p. 97) arose. It is possible to define a linear function without using components: l is linear if

$$l(\xi + \eta) = l(\xi) + l(\eta) \qquad \text{and} \qquad l(\alpha \cdot \xi) = \alpha \cdot l(\xi)$$

for every ξ and η in its domain and every real number α. Then we define the Fréchet derivative of f at α to be the linear function l for which

$$\frac{|f(\alpha + \delta) - f(\alpha) - l(\delta)|}{|\delta|} \to 0$$

as $\delta \to 0$.

A third development is the study of functions on manifolds. A manifold in $\mathbb{R}^2$ is a curve, a manifold in $\mathbb{R}^3$ is a curve or a surface, and so on. Manifolds occur naturally in physics. For example, if we let $\mathbb{D}$ be the set of all trios $(\alpha_1, \alpha_2, \alpha_3)$ for which a given specimen of gas can have simultaneously pressure α_1, volume α_2, and temperature α_3, then $\mathbb{D}$ is a manifold in $\mathbb{R}^3$. Each element of $\mathbb{D}$ is called a 'state' of the specimen, and any 'function of state' is a function with domain $\mathbb{D}$.

It turns out that manifolds are the appropriate setting for much of the calculus used in applied mathematics.

Answers and hints to exercises

Chapter 1

1. $\mathbb{R}^3$; yes, $f^{\sim}(u, v, w) = (\frac{1}{2}(u - v + w),$
   $\frac{1}{2}(u + v - w), \frac{1}{2}(-u + v + w))$ with domain $\mathbb{R}^3$.
2. $\{(u, v, w): u + v + w = 0\}$; no.
3. $f(\mathbb{A}) = \{(x, y, z): x < y + z, \quad y < z + x \quad$ and $\quad z < x + y\}$;
   $g(\mathbb{A}) = \operatorname{rng} g$;
   $f^{\sim}(\mathbb{A}) = \{(x, y, z): x + y > 0, \quad y + z > 0 \quad$ and $\quad z + x > 0\}$;
   $g^{\sim}\mathbb{A} = \varnothing$;
   $f \circ g(x, y, z) = (x - z, y - x, z - y)$ with domain $\mathbb{R}^3$;
   $g \circ f = f \circ g$;
   $f \circ f(x, y, z) = (x + 2y + z, x + y + 2z, 2x + y + z)$ with
   domain $\mathbb{R}^3$;
   $g \circ g(x, y, z) = (x - 2y + z, y - 2z + x, z - 2x + y)$ with
   domain $\mathbb{R}^3$.
4. $g \circ f(x, y) = (2x + 3y, -2x - 3y)$ with domain $\mathbb{R}^2$;
   $f \circ g(x) = -x$ with domain $\mathbb{R}$; $f \circ f$ and $g \circ g$ have empty
   domains.
5. Values at x are $x, -x, x^4 + 2x^2 + 1, x^2 + x - 1, x, 2x^2$.
6. (a) $(4, 6)$ (b) (a, b, c) (c) $(-6, 4)$ (d) $(-1, -1, 2)$
   (e) 2   (f) 3   (g) and (h) do not exist.
7. $(f + g)(x, y, z) = (2x, 2y)$; $(2f - 3g)(x, y, z) = (5y - x,$
   $5z - y)$;
   $(f \cdot g)(x, y, z) = x^2 - z^2$; $(f + (1, 0))(x, y, z) = (x + y + 1,$
   $y + z)$; all with domain $\mathbb{R}^3$.
8. (a) $f(\cdot, 0) = t$, $f(1, \cdot) = 1 + 2t$, both with domain $\mathbb{R}$.
   (b) $f(1, \cdot, \cdot) = v + u^2$ with domain $\mathbb{R}^2$; $f(1, 2, \cdot) = t + 4$
   with domain $\mathbb{R}$.
   (c) Yes.

Chapter 2

1. It is a triangle; no.
2. It consists of two squares.
3. 8.
4. Hint: consider $\{x: (x, y) \in \mathbb{S}$ for some $y\} \times \{y: (x, y) \in \mathbb{S}$
   for some $x\}$.

5. Hint: if $(x^2 + y^2)^{1/2} < k$, consider a neighbourhood of (x, y) whose side is between $(x^2 + y^2)^{1/2}$ and k.

6. $[0; 1] \times [0; 1] \times [0; 1]$.

7. Obviously $\mathbb{S}$ is contained in the intersection. If $x \notin \mathbb{S}$, the complement of $\{x\}$ is an open set containing $\mathbb{S}$, and x does not belong to it. Therefore x does not belong to the intersection.

8. $(4/5, -3/5)$ or $(-4/5, 3/5)$; two.

9. By the triangle inequality $|\alpha - \beta| + |\beta| \geq |(\alpha - \beta) + \beta|$, whence $|\alpha - \beta| \geq |\alpha| - |\beta|$; also $|\alpha| + |-\beta| \geq |\alpha + (-\beta)|$, whence $|\alpha - \beta| \leq |\alpha| + |\beta|$.

10. By the proof of the first inequality for the norm (p. 13), $|\alpha| \cdot |\beta| = \alpha \cdot \beta$ if and only if $|\alpha| \beta = |\beta| \alpha$. $|\alpha + \beta| = |\alpha| + |\beta|$ under the same condition. $|\alpha| + |\beta| = |\alpha - \beta|$ if $|\alpha| = 0$ or $\beta = x\alpha$ with $x \geq 0$. $|\alpha| - |\beta| = |\alpha - \beta|$ if $|\beta| = 0$ or $\alpha = x\beta$ with $x \geq 1$.

11. Hint: the set is the difference between two open discs.

12. Hint: if $|\alpha - \alpha^*| < r + r^*$ the discs have $\alpha^* + k(\alpha - \alpha^*)/|\alpha - \alpha^*|$ in common where k is any number between $|\alpha - \alpha^*| - r$ and r^*. If $|\alpha - \alpha^*| > r + r^*$, then $|\xi - \xi^*| \geq |\alpha - \alpha^*| - |\xi - \alpha| - |\xi^* - \alpha^*|$, giving $\inf |\xi - \xi^*| \geq |\alpha - \alpha^*| - r - r^*$. For the reverse inequality, consider $\alpha + (r - k)(\alpha^* - \alpha)/|\alpha^* - \alpha|$ and $\alpha^* + (r^* - k)(\alpha - \alpha^*)/|\alpha - \alpha^*|$ for small positive k. The answer to the second part is $|\beta|$, where $\beta_i = \max(0, |\alpha_i - \alpha_i^*| - r - r^*)$.

13. True.

14. All except the first.

15. $[0; 1]$; $\varnothing$.

16. $\{(x, y) : x = 0 \text{ and } y \geq 0, \text{ or } y = 0 \text{ and } x \geq 0\}$; $\{(x, y) : x > 0 \text{ and } y > 0\}$; the rest of $\mathbb{R}^2$.

17. Hint: consider the open disc of radius 1 and centre η, where $|\eta| > 1 + \sup\{|\xi| : \xi \in \mathbb{S}\}$.

18. If $\mathbb{S}$ is open, any member belongs to int $\mathbb{S}$ and so does not belong to bdy $\mathbb{S}$. If $\mathbb{S}$ contains no boundary points its members (which cannot belong to ext $\mathbb{S}$) all belong to int $\mathbb{S}$.

19. No; no; bdy$(\mathbb{U} \cup \mathbb{V}) \subseteq$ bdy $\mathbb{U} \cup$ bdy $\mathbb{V}$ (hint: a point interior to $\mathbb{U}$ cannot lie in bdy$(\mathbb{U} \cup \mathbb{V})$); no (consider the set of all $\{x\}$ for rational x); bdy$(\mathbb{U} \cap \mathbb{V}) \subseteq$ bdy $\mathbb{U} \cup$ bdy $\mathbb{V}$.

20. Yes.

21. Yes; no.

22. No.

23. Yes; yes.

Chapter 3

1. Yes (hint: $|x^3/(x^2 + y^2)| \leq |(x, y)|$).
2. Yes.
3. No
4. No.
6. Yes, $c = 0$.
7. f is a component of the identity.
8. Note that $g(x, y) = xy \cdot f(x, y)$.
9. Wherever $x^2 \neq y^2$.
10. Hint: consider $|f(\xi) - f(\alpha)|$ and $|f(\xi) - g(\alpha)|$.
11. Hint: if $(f^{\frown}\mathbb{W}) \cap \mathbb{A} = \mathbb{V} \cap \mathbb{A}$, then $(f^{\frown}\mathbb{W}) \cap \mathbb{A} \cap \mathbb{B} = \mathbb{V} \cap \mathbb{A} \cap \mathbb{B}$, i.e. $(f^{\frown}\mathbb{W}) \cap \mathbb{B} = \mathbb{V} \cap \mathbb{B}$.
12. $f^{\frown}\mathbb{N}$ contains the neighbourhood $\mathbb{U}$ mentioned in the definition of continuity.
13. If x is interior to $[a; b]$ then f is continuous at x if and only if it is continuous on $[a; b]$ at x. If f is continuous at $a+$ then for each neighbourhood $\mathbb{N}$ of $f(a)$ there is a neighbourhood $\mathbb{U}$ of a such that $f(a) \in \mathbb{N}$ whenever $x \in \mathbb{U}$ and $x \geq a$. If $x \in \mathbb{U} \cap [a; b]$ then $x \in \mathbb{U}$ and $x \geq a$. Therefore $f(a) \in \mathbb{N}$ whenever $x \in \mathbb{U} \cap [a; b]$. Consequently f is continuous on $[a; b]$ at a. A similar argument holds for $b-$ and for the converse.
14. No, e.g. $f(x) = x^2$, $\mathbb{U} = (-1; 1)$.
15. Yes; let g be the restriction of f to $f^{\frown}\mathbb{U}$ and then dom g is open and so $g^{\frown}\mathbb{U}$ (which is the same as $f^{\frown}\mathbb{U}$) is open.
16. Yes.
17. There may not be a pair of nearest points, e.g. $\mathbb{A} = \{(x, 1/x) : x > 0\}$, $\mathbb{B} = \{(x, -1/x) : x > 0\}$.
18. Yes; yes.
19. dom $f = \{(x, 1) : x \geq 0\} \cup \{(x, -1) : x < 0\}$, $f(x, 1) = x$ if $x \geq 0$, $f(x, -1) = x$ if $x < 0$.
    dom $f = \{(x, y) : y > 0$ and $0 \leq x < 2\pi\}$, $f(x, y) = (y \cos x, y \sin x)$.
20. dom $f = \{x : x \geq 1\}$, $f(x) = (\cos(2\pi/x), \sin(2\pi/x))$.
21. $f^{1/2}$, f and f^3.
22. $f^{1/2}$ and f.
23. $|(x^2 + 1)^{-1} - (y^2 + 1)^{-1}| = |x^2 - y^2| (x^2 + 1)^{-1}(y^2 + 1)^{-1}$

$$= |x - y| |x + y| (x^2 + 1)^{-1}(y^2 + 1)^{-1}$$

$$\leq |x - y| \left(\left| \frac{x}{x^2 + 1} \right| + \left| \frac{y}{y^2 + 1} \right| \right) \leq |x - y|.$$

Therefore $|f(x) - f(y)| < \delta$ whenever $|x - y| < \delta$ and so ϵ itself will do for δ.

24. f and f^2.

25. No.

26. Let $\mathbb{A}$ be the subset. Any property that holds for every (x, y) in $\mathbb{R}^2$ holds for every (x, y) in $\mathbb{A}$.

27. Follow the pattern of the corresponding theorem for ordinary continuity. (The point of the exercise is for you to see for yourself that you can do so.)

28. Let $\mathbb{A}$ be the first open disc and $\mathbb{B}$ the second. $\mathbb{A}$ and $\mathbb{B}$ are both open in $\mathbb{S}$.

29. Any two points can be joined by a sequence of line segments. The full details of the proof of this fact may be tedious.

30. If $\mathbb{S}$ is connected and split into proper subsets $\mathbb{A}$ and $\mathbb{B}$, let α be a point of $\mathbb{A}$ that does not have a neighbourhood $\mathbb{U}_\alpha$ for which $\mathbb{S} \cap \mathbb{U}_\alpha \subseteq \mathbb{A}$. Then every neighbourhood of α contains both a point of $\mathbb{B}$ and a point (namely α) of $\mathbb{A}$. Therefore α is a boundary point of $\mathbb{B}$. Conversely, if $\mathbb{S}$ is disconnected it can be split into proper subsets $\mathbb{A}$ and $\mathbb{B}$ with the property that every point α of $\mathbb{A}$ has a neighbourhood $\mathbb{U}_\alpha$ such that $\mathbb{S} \cap \mathbb{U}_\alpha \subseteq \mathbb{A}$. Then $\mathbb{U}_\alpha$ cannot contain a point of $\mathbb{B}$, and so α is not a boundary point of $\mathbb{B}$. Similarly, no point of $\mathbb{B}$ is a boundary point of $\mathbb{A}$.

31. No subset of $\mathbb{S}$ open in $\mathbb{S}$ can contain any point of $\{(x, \sin(1/x)) : x > 0\}$ without containing all this set. Similarly for $\{(0, x) : x \in \mathbb{R}\}$. Thus the only split of $\mathbb{S}$ into proper subsets open in $\mathbb{S}$ is into these two sets. But every point of the second is a boundary point of the first.

32. No; see exercise 28.

33. If $\mathbb{P}$ and $\mathbb{Q}$ exist, $\mathbb{S} \cap \mathbb{P}$ and $\mathbb{S} \cap \mathbb{Q}$ are open in $\mathbb{S}$. If $\mathbb{S}$ splits into $\mathbb{A}$ and $\mathbb{B}$, let $\mathbb{P}$ be the union of all $\mathbb{U}_\alpha$ for α in $\mathbb{A}$, and $\mathbb{Q}$ the union of all $\mathbb{U}_\alpha$ for α in $\mathbb{B}$.

34. If $\mathbb{T}$ splits into proper subsets $\mathbb{A}$ and $\mathbb{B}$ open in $\mathbb{T}$ then $\mathbb{S}$ splits into $\mathbb{S} \cap \mathbb{A}$ and $\mathbb{S} \cap \mathbb{B}$, which can easily be proved to be open in $\mathbb{S}$. Then one of these, say $\mathbb{S} \cap \mathbb{A}$, is $\mathbb{S}$ and the other is empty. Let $\beta \in \mathbb{B}$. It has a neighbourhood $\mathbb{U}_\beta$ such that $\mathbb{T} \cap \mathbb{U}_\beta \subseteq \mathbb{B}$. But $\beta \notin \mathbb{S}$ (because $\mathbb{S} \cap \mathbb{B}$ is empty) and is therefore a boundary point of $\mathbb{S}$. Consequently $\mathbb{U}_\beta$ contains a point belonging to $\mathbb{S}$ (and therefore to $\mathbb{T}$) but not to $\mathbb{B}$. This contradicts the fact that $\mathbb{T} \cap \mathbb{U}_\beta \subseteq \mathbb{B}$.

35. If $\mathbb{S}$ is a connected subset of $\mathbb{R}$, b and c belong to $\mathbb{S}$, and

k is between b and c, then $k \in \mathbb{S}$ because if not $\mathbb{S}$ splits into proper subsets $\{x \in \mathbb{S} : x < k\}$ and $\{x \in \mathbb{S} : x > k\}$ open in $\mathbb{S}$. For the second part see exercise 29.

36. The limit fails to exist for (a) and (b); it is zero for (c) and (d).

37. (a) $\lim_{(0,0)} f$, $\lim_{x \to 0} [\lim_0 f(x, \cdot)]$ and $\lim_{y \to 0} [\lim_0 f(\cdot, y)]$ fail to exist. The other two are zero.
    (b) $\lim_{(0,0)} f$ fails to exist. The others are zero.
    (c) $\lim_{(0,0)} f$ fails to exist. $\lim_0 f(\cdot, 0) = \lim_{x \to 0} [\lim_0 f(x, \cdot)]$ $= 1$. The other two are -1.

38. No; see exercise 37(b).

Chapter 4

1. $3x^2 y^4$; $4x^3 y^3$.
2. $D_1 f(x, y) = 2x$, $\qquad D_2 f(x, y) = -2y$, $\qquad D_1 f(a, b) = 2a$, $D_2 f(x, x) = -2x$, $D_1 f(y, x) = 2y$.
3. $D_1 f(1, 2, 3) = 6$, $D_2 f(1, 2, 3) = 3$, $D_3 f(1, 2, 3) = 2$.
4. No.
5. $D_1 f(x, y) = -y/(x^2 + y^2)$, $D_2 f(x, y) = x/(x^2 + y^2)$.
6. $\lim_{h \to 0} [f(x, y + h, z) - f(x, y, z)]/h$.
7. $f(y, x) = -f(x, y)$. Therefore $\lim_{h \to 0} [f(x + h, y) - f(x, y)]/h = -\lim_{h \to 0} [f(y, x + h) - f(y, x)]/h$.
8. 0.
9. $f(u, v, w)' = D_1 f(u, v, w) u' + D_2 f(u, v, w) v'$
    $\qquad\qquad + D_3 f(u, v, w) w'$.
10. 0.
11. $q' = D_1 f(\sin, \cos)\cos - D_2 f(\sin, \cos)\sin$.
12. $D_1[f(u, v)] = D_1 f(u, v) D_1 u + D_2 f(u, v) D_1 v$,
    $D_2[f(u, v)] = D_1 f(u, v) D_2 u + D_2 f(u, v) D_2 v$.
13. The first is $D_1 v = f'(u) D_1 u$, $D_2 v = f'(u) D_2 u$ and $D_3 v = f'(u) D_3 u$.
    The second is $v' = D_1 f(u_1, u_2, u_3) u_1' + D_2 f(u_1, u_2, u_3) u_2'$
    $\qquad\qquad + D_3 f(u_1, u_2, u_3) u_3'$.
    The third is $D_1 v_1 = D_1 f_1(u_1, u_2) D_1 u_1 + D_2 f_1(u_1, u_2) D_1 u_2$
    $\qquad\quad D_1 v_2 = D_1 f_2(u_1, u_2) D_1 u_1 + D_2 f_2(u_1, u_2) D_1 u_2$
    $\qquad\quad D_2 v_1 = D_1 f_1(u_1, u_2) D_2 u_1 + D_2 f_1(u_1, u_2) D_2 u_2$
    $\qquad\quad D_2 v_2 = D_1 f_2(u_1, u_2) D_2 u_1 + D_2 f_2(u_1, u_2) D_2 u_2$.
14. $D_1 f_1 = 2x \sin z$, $D_1 f_2 = 0$, $D_2 f_1 = 0$, $D_2 f_2 = -z \sin(yz)$, $D_3 f_1 = x^2 \cos z$, $D_3 f_2 = -y \sin(yz)$.

15. $D_1g(r, s) = D_1f(r\cos s, r\sin s)\cos s$
$\qquad + D_2f(r\cos s, r\sin s)\sin s$
$D_2g(r, s) = -D_1f(r\cos s, r\sin s)r\sin s$
$\qquad + D_2f(r\cos s, r\sin s)r\cos s$.

16. $D_1g(x + y, x - y) = 0$; $f(x, y) = F(x - y)$ where F is a differentiable function in $\mathbb{R}$ into $\mathbb{R}$.

17. $D_1^2f(x, y) =$
$\qquad (4D_1^2g + 2yD_1D_2g + 2yD_2D_1g + y^2D_2^2g)(2x + 3y, xy)$
$D_2^2f(x, y) =$
$\qquad (9D_1^2g + 3xD_1D_2g + 3xD_2D_1g + x^2D_2^2g)(2x + 3y, xy)$
$D_2D_1f(x, y) =$
$\qquad (6D_1^2g + 3yD_1D_2g + 2xD_2D_1g + xyD_2^2g)(2x + 3y, xy)$.

18. $x^2D_1^2g(u, v) = 0$.

19. $f(x, y) = xF(y/x) + G(y/x)$ where F and G are differentiable functions in $\mathbb{R}$ into $\mathbb{R}$.

20. $\partial q/\partial u = \cos v + v^2\cos u$, $\qquad \partial q/\partial v = -u\sin v + 2v\sin u$

21. $\partial r/\partial x = x/(x^2 + y^2)^{1/2}$, $\qquad \partial r/\partial y = y/(x^2 + y^2)^{1/2}$,
$\partial\theta/\partial x = -y/(x^2 + y^2)$, $\qquad \partial\theta/\partial y = x/(x^2 + y^2)$.

22. $\partial x/\partial r = \cos\theta$, $\qquad \partial x/\partial\theta = -r\sin\theta$,
$\partial y/\partial r = \sin\theta$, $\qquad \partial y/\partial\theta = r\cos\theta$.

23. (a) $\partial q/\partial x = 2x - 2y$.
(b) $q = 3x^2 - 2xu$, $\qquad \partial q/\partial x = 6x - 2u$.
(c) $q = 5x^2/2 - xv/2$, $\qquad \partial q/\partial x = 5x - v/2$.

24. $(\partial w/\partial u)_v = (\partial w/\partial x)_y \cos a + (\partial w/\partial y)_x \sin a$
$(\partial w/\partial v)_u = (\partial w/\partial x)_y \sin a - (\partial w/\partial y)_x \cos a$
$(\partial w/\partial x)_y = (\partial w/\partial u)_v \cos a + (\partial w/\partial v)_u \sin a$
$(\partial w/\partial y)_x = (\partial w/\partial u)_v \sin a - (\partial w/\partial v)_u \cos a$.

25. $(\partial u/\partial p)_q = 2p$, $\quad (\partial u/\partial q)_p = -2q$, $\quad (\partial v/\partial p)_q = 2q$,
$(\partial v/\partial q)_p = 2p$, $\quad (\partial p/\partial v)_u = q/2(p^2 + q^2)$, $\quad (\partial q/\partial u)_v =$
$-q/2(p^2 + q^2)$, $\quad (\partial p/\partial u)_v = (\partial q/\partial v)_u = p/2(p^2 + q^2)$.

26. $\partial w/\partial u = \{(\partial w/\partial x)_y(\partial v/\partial y)_x - (\partial w/\partial y)_x(\partial v/\partial x)_y\}/$
$\qquad \{(\partial u/\partial x)_y(\partial v/\partial y)_x - (\partial u/\partial y)_x(\partial v/\partial x)_y\}$

27. Hint: $(\partial w/\partial x)_y = \partial w/\partial x + (\partial w/\partial u)(\partial u/\partial x)_y$
$\qquad + (\partial w/\partial v)(\partial v/\partial x)_y$ and
$(\partial v/\partial x)_y = \partial v/\partial x + (\partial v/\partial u)(\partial u/\partial x)_y$.
$(\partial w/\partial y)_x = (\partial w/\partial u)(\partial u/\partial y)_x + (\partial w/\partial v)(\partial v/\partial y)_x$
$\qquad + (\partial w/\partial v)(\partial v/\partial u)(\partial u/\partial y)_x$.

28. $\partial p/\partial u$ with respect to $p = f(u, v, g(u, v))$ is equal to $\partial p/\partial u + (\partial p/\partial w)(\partial w/\partial u)$ where these three derivatives are with respect to (15). Alternatively $(\partial p/\partial u)_v = (\partial p/\partial u)_{v,w} + (\partial p/\partial w)_{u,v}(\partial w/\partial u)_v$.

29. (a) -1.
(b) -1.

30. Yes, provided that they are defined with respect to equivalent relations.

31. $-r \sin \theta$, $\cos \theta$.

32. $\partial w / \partial r = (\partial w / \partial x) \cos \theta + (\partial w / \partial y) \sin \theta$. Find a similar formula for $\partial w / \partial \theta$ and compute the right-hand side of the desired relation.

33. $\partial^2 q / \partial u^2 = -\partial^2 q / \partial v^2 = 2uv/(u^2 + v^2)^2$; $\partial^2 q / \partial u \, \partial v = (v^2 - u^2)/(u^2 + v^2)^2$.

34. For any function w in $\mathbb{R}^2$ into $\mathbb{R}$, $\partial w / \partial x = \cos \theta \cdot \partial w / \partial r - r^{-1} \sin \theta \cdot \partial w / \partial \theta$. Set $w = q$ in this equation and substitute the resulting formula for $\partial q / \partial x$ into the right-hand side. After simplifying, we obtain

$$\frac{\partial^2 q}{\partial x^2} = \cos^2 \theta \frac{\partial^2 q}{\partial r^2} - 2r^{-1} \sin \theta \cos \theta \frac{\partial^2 q}{\partial r \, \partial \theta}$$

$$+ r^{-2} \sin^2 \theta \frac{\partial^2 q}{\partial \theta^2} + r^{-1} \sin^2 \theta \frac{\partial q}{\partial r} + 2r^{-2} \sin \theta \cos \theta \frac{\partial q}{\partial \theta}.$$

Similarly,

$$\frac{\partial^2 q}{\partial x \, \partial y} = \sin \theta \cos \theta \frac{\partial^2 q}{\partial r^2} + r^{-1}(\cos^2 \theta - \sin^2 \theta) \frac{\partial^2 q}{\partial r \, \partial \theta}$$

$$- r^{-2} \cos \theta \sin \theta \frac{\partial^2 q}{\partial \theta^2} - r^{-1} \cos \theta \sin \theta \frac{\partial q}{\partial r}$$

$$+ r^{-2}(\sin^2 \theta - \cos^2 \theta) \frac{\partial q}{\partial \theta}$$

and

$$\frac{\partial^2 q}{\partial y^2} = \sin^2 \theta \frac{\partial^2 q}{\partial r^2} + 2r^{-1} \cos \theta \sin \theta \frac{\partial^2 q}{\partial r \, \partial \theta} + r^{-2} \cos^2 \theta \frac{\partial^2 q}{\partial \theta^2}$$

$$+ r^{-1} \cos^2 \theta \frac{\partial q}{\partial r} - 2r^{-2} \cos \theta \sin \theta \frac{\partial q}{\partial \theta}.$$

35. Each equals $a^2 f''(u + av) + a^2 g''(u - av)$.

36. Use the technique of exercise 34.

37. For any function w of x, y, and u, $(\partial w / \partial x)_y = \partial w / \partial x + (\partial w / \partial u)(\partial u / \partial x)_y$; use this with $w = v$, with $w = (\partial v / \partial x)_y$, and with $w = (\partial u / \partial x)_y$.

38. $\mathrm{D}u = u'$ and $\mathrm{D}v = v'$, and so $u \cdot \mathrm{D}v + v \cdot \mathrm{D}u = \mathrm{D}(uv) = \kappa \mathrm{D}w$. The formula still holds in higher dimensions if we change uv to $u \cdot v$.

39. $p(\xi)D_1u(\xi) + q(\xi)D_1v(\xi) = 0$ and $p(\xi)D_2u(\xi) + q(\xi)D_2v(\xi) = 0$ for every ξ. If the determinant of this system of equations is non-zero, the only solution is $p(\xi) = q(\xi) = 0$.

40. Each equals $D_1f(u, v)u' + D_2f(u, v)v'$.

41. $f(\alpha_1 + \delta_1, \alpha_2 + \delta_2)$

$$= f(\alpha_1, \alpha_2) + \delta_1 D_1 f(\alpha_1 + \theta\delta_1, \alpha_2 + \theta\delta_2)$$
$$+ \delta_2 D_2 f(\alpha_1 + \theta\delta_1, \alpha_2 + \theta\delta_2)$$
or $f(\alpha_1, \alpha_2) + \delta_1 D_1 f(\alpha_1, \alpha_2) + \delta_2 D_2 f(\alpha_1, \alpha_2)$
$$+ \tfrac{1}{2}\delta_1{}^2 D_1{}^2 f(\alpha_1 + \theta\delta_1, \alpha_2 + \theta\delta_2)$$
$$+ \delta_1\delta_2 D_1 D_2 f(\alpha_1 + \theta\delta_1, \alpha_2 + \theta\delta_2)$$
$$+ \tfrac{1}{2}\delta_2{}^2 D_2{}^2 f(\alpha_1 + \theta\delta_1, \alpha_2 + \theta\delta_2).$$

42. $\delta_1{}^3 D_1{}^3 g(0, 0) + 3\delta_1{}^2\delta_2 D_1{}^2 D_2 g(0, 0)$
$$+ 3\delta_1\delta_2{}^2 D_1 D_2{}^2 g(0, 0) + \delta_2{}^3 D_2{}^3 g(0, 0).$$

43. $|f(\delta)| \le \tfrac{1}{2}\kappa(|\delta_1| + |\delta_2|)^2$.

44. Use Taylor's theorem, replacing δ_1 by $\cos\theta$, δ_2 by $\sin\theta$, and θ by x.

45. (b), (d).

46. x is a function of y; z is a function of x and y.

47. Yes, except for (c) at $(0, 0)$.

48. They are equal on $\mathbb{U} \cap \mathbb{V}$.

49. If $(x, y) \in \mathbb{U} \cap \mathbb{W}$ and $f(x, y) = 0$, then $q(x) = s(x)$.

50. Set $f(x, y) = x^5 + y^5 + xy - 3$ for every x and y of $\mathbb{R}$. Then f is continuously differentiable, $f(1, 1) = 0$, and $D_2f(1, 1) \ne 0$.

51. f is continuously differentiable on a neighbourhood of (a, b). $D_2f(a, b) = 4b(a^2 + b^2 + 1) \ne 0$; $q'(a) = -a(a^2 + b^2 - 1)/(a^2 + b^2 + 1)b$.

52. The tangent is horizontal at $(\pm\tfrac{1}{2}\sqrt{3}, \pm\tfrac{1}{2})$; it is vertical at $(\pm\sqrt{2}, 0)$. (At $(0, 0)$ the curve crosses itself.)

53. $f(x, y) = x - y^2$; $a = b = -c = 1$ is one possibility.

54. No.

55. $f'(1) \ne 0$.

56. At points $(0, t, -t)$ and $(t, 0, -t)$ for every non-zero number t.

57. $0 : -2 : 1$.

58. At points $(x^2 + y^2, 2xy)$ for which $x^2 \ne y^2$.

59. All except $(0, 0)$.

60. (a), (b) No. (c) Yes (on any neighbourhood in the domain of f).

61. (a) Minimum at $(0, 0)$.
    (b) Maximum at $(\pi/3, \pi/3)$, minimum at $(-\pi/3, -\pi/3)$ modulo 2π.
    (c) Maximum at $(-\frac{1}{2}, 4)$.
    (d) Minimum at $(1, 1)$.
    (e) Maximum (non-strict) where $x = y \neq 0$; minimum (non-strict) where $x = -y \neq 0$.
    (f) Maximum at $(x, 0)$ for $x > 6$ or $x < 0$, and at $(3, 2)$; minimum at $(x, 0)$ for $0 < x < 6$.
62. $f(x, y) = -(x^2 + y^2)^2$, $\alpha = (0, 0)$; $f(x, y) = x^3 y^3$, $\alpha = (0, 0)$; it must be zero.
63. $\alpha = (0, 0)$, $f(x, y) = -x^2$ if $y = 0$, $f(x, y) = y^2$ if $x = 0$, $f(x, y) = 0$ otherwise.
64. Let $\mathbb{S} = \{(x, y) : x \geq 2/3 \text{ and } y \geq 1/3 \text{ and } xy \leq 6\}$; 12 at $(2, 1)$.
65. Yes.
66. No.
67. Function continuous on compact set.
68. $(a + c \pm \{(a - c)^2 + 4b^2\}^{1/2}/2(ac - b^2)$ if $b^2 < ac$; $(a + c - \{(a - c)^2 + 4b^2)\}^{1/2}/2(ac - b^2)$ if $b^2 > ac$.
69. 8 and 20/3.
70. The tangents at $(1, 1)$ and $(-1, -1)$.
71. No.
72. $8|abc|/3\sqrt{3}$.
73. The greatest and least of $a^2 p^2$, $b^2 q^2$, $c^2 r^2$.
74. Greatest, -6; no least.
75. 36 and 1.
76. 2 and 18/35.
77. (a) nearest $(-1, 0, 1)$, no furthest.
    (b) $(-1, 0, 1)$ and $(3, 0, 3)$.
78. $\pm 1/\sqrt{2}$.
79. 17/27 and 1.
80. $\pm 1/\sqrt{6}$.

Chapter 5

1. (a) Each cell $\mathbb{Q}_r$ of $\mathcal{P}_n$ is a union of cells $\mathbb{Q}_i$ of $\mathcal{P}_p$, and $\sup f(\mathbb{Q}_r) \geq \sup f(\mathbb{Q}_i)$ for each of these $\mathbb{Q}_i$. Then $m_r c(\mathbb{Q}_r) \geq \sum m_i c(\mathbb{Q}_i)$. Summing over r gives the result.
   (b) No.
2. Yes, no, yes.
3. $\frac{1}{2}(1 + 1/n)(2 + 1/n)$, $\frac{1}{2}(1 - 1/n)(2 - 1/n)$, 1, 1.

4. Yes.
5. Yes.
6. Clearly, g is bounded. Let n be the number of points at which f and g are unequal and $k = \sup |f - g|$. If ϵ is any positive number, choose a partition $\mathscr{P}$ of mesh less than ϵ / kn. Then $^*\Sigma_{\mathscr{P}} g \leq \int f + \epsilon$, whence $^*\int g \leq \int f$. Similarly $_*\int g \geq \int f$.
7. Hint: $\sup_Q |f| - \inf_Q |f| \leq \sup_Q - \inf_Q f$. Now compare $^*\Sigma_{\mathscr{P}} |f| - _*\Sigma_{\mathscr{P}} |f|$ with $^*\Sigma_{\mathscr{P}} f - _*\Sigma_{\mathscr{P}} f$.
8. Hint: every cell of a partition of $\mathbb{J}$ is a cell of some partition of dom f.
9. $4/3$.
10. Yes.
11, 12, 13. Apply the results of the text to $(f + g)_{\mathbb{A}}|_{\mathbb{I}}$, etc.
14. Yes, $\int_{\mathbb{A}} f = 1$.
15. Yes; yes.
16. No (the boundary may not be negligible).
17. $\frac{1}{2}ac$ (integrate the function that takes the value 1 on the set and the value 0 off the set).
18. The integrand is bounded and continuous, and the boundary is negligible.
19. π; πr^2.
20. $f_{\mathbb{A}} = c\mathbf{1}_{\mathbb{A}}$, where c is the value of f.
21. $\left| \det \begin{vmatrix} \kappa_1 & \kappa_2 \\ \lambda_1 & \lambda_2 \end{vmatrix} \right|$.
22. $\mathbf{1}_{\mathbb{B}}(\tau) \leq \mathbf{1}_{\mathbb{A}}(\tau)$; $c(\mathbb{B}) \leq c(\mathbb{A})$ if both exist.
23. $\frac{1}{2}$.
24. 1.
25. Corners at $(1, 0)$, $(1.1, -2.1)$, $(1.1, 2.1)$, $(1.21, 0)$; sides slightly curved.
26. If $f(x, y) = (ax + by + p, \ cx + dy + q)$, then $Jf(x, y) = \left| \det \begin{vmatrix} a & b \\ c & d \end{vmatrix} \right|$. The converse is far from true.
27. Yes.
28. Find a sequence of rectangles exhausting $\mathscr{F}$ and apply exercise 27.
29. $31a/6 + 2b/3 + 9c/4 + 3d/4$.
30. Set $\mathbb{B} = \{(x, y) : x \geq 0, \ y \geq 0, \ 3 \leq x \leq 4 \text{ and } 1 \leq y \leq 2\}$, and $g(x, y) = (x^2 + 2y^2, \ x^2 - 3y^2)$. Then $g(\mathbb{A}) = \mathbb{B}$, g is one-to-one between $\mathbb{A}$ and $\mathbb{B}$, and $\mathbb{B}$ is a set over which it is easy to integrate. For any integrable h, $\int_{\mathbb{A}} h \circ g \cdot Jg = \int_{\mathbb{B}} h$, so we want to choose h so that $h \circ g \cdot Jg = f$, i.e.

$h(x^2 + 2y^2, x^2 - 3y^2) |20xy| = xy$, i.e. h is constant with value $1/20$. Then $\int_{\mathbb{B}} h = c(\mathbb{B})/20 = 1/20$.

31. Zero, by symmetry.

32. $-1/20$.

33. $3(2^{n+1} - 1)/4(n + 1)$ if $n \neq -1$, $\frac{3}{2} \ln 2$ if $n = -1$.

34. $\frac{1}{2} \sin 1$.

35. $f(x) = x^{-1/2}$ with domain $(0; 1)$; $g(x) = 1/x$ with domain $\{x : x > 1\}$.

36. $f(x) = (x^2 + x^{1/2})^{-1}$ with domain $\{x : x > 0\}$, $g(x) = \tan(x^2)$ with domain $(0, (\pi/2)^{1/2})$.

37. $\int_{\mathbb{A}} f = (m + 1)^{-1} b_1 b_2 \ldots b_n (b_1{}^m + b_2{}^m + \ldots + b_n{}^m)$ if $m \geq 0$, and does not exist if $m < 0$.
$\int_{\mathbb{A}} g = 2^{-n} b_1 b_2 \ldots b_n$.

38. $a + b + c$.

Answers and hints to some of the problems

Chapter 1

2. No, yes if $\mathbb{A} \subseteq \operatorname{rng} f$
3. Try $f = \sin$.

Chapter 3

4. Yes.
5. Let $g(x) = (x_1, \ldots, x_m, f_{m+1}(x), \ldots, f_n(x))$ where the f_i are continuous. Then $g(\operatorname{dom} f)$ is compact if and only if $\operatorname{dom} f$ is.
6. $\mathbb{S} = \mathbb{R}$, $f = \operatorname{arc} \tan$.
7. (a) $\operatorname{dom} f = (-2, -1) \cup (0, 1]$, $f(x) = x^2$, $\mathbb{S} = (0, 4)$.
   (b) Use a function in $\mathbb{R}$ into $\mathbb{R}^2$.
8. Let ξ be any member of $\operatorname{bdy} f(\mathbb{S})$. Then $\xi \in \operatorname{rng} f$ and so $\xi = f(f^{\sim}(\xi))$. If $\mathbb{U}$ is any neighbourhood of $f^{\sim}(\xi)$, $f(\mathbb{U})$ is open (because $f^{\sim}$ is continuous) and contains a neighbourhood of ξ, which contains a point α in $f(\mathbb{S})$ and a point β not in $f(\mathbb{S})$. $f^{\sim}(\alpha)$ belongs to $\mathbb{S}$ but $f^{\sim}(\beta)$ does not; consequently $f^{\sim}(\xi) \in \operatorname{bdy} \mathbb{S}$ and $\xi \in f(\operatorname{bdy} \mathbb{S})$. For the reverse implication, interchange f with $f^{\sim}$.
20. No: e.g. $\mathbb{S} = (-1; 0)$, $\mathbb{T} = (0; 1)$, $f(x) = 0$ if $-1 < x < 0$, $f(x) = 1$ if $0 < x < 1$.

Chapter 4

1. No; no.
2. Yes.
3. Hint: evaluate $D_1 f(0, 0)$ and $D_1 f(x, y)$ separately for $(x, y) \neq 0$.
5. The first three answers are 'yes'.

6. On any interval that does not have an odd multiple of $\frac{1}{2}\pi$ in its interior.

8. No, e.g. $f(x, y) = 3xe^y - x^3 - e^{3y}$.

9. No, e.g. $f(x, y) = 2y(x^2 + \frac{3}{2}) - x^4 - 2x^2$ if $\frac{1}{2}x^2 < y < x^2$, $f(x, y) = x^2 - 3y$ if $0 < y \le \frac{1}{2}x^2$, $f(x, y) = x^2 + y^2$ otherwise.

10, 11. Hint: be careful about the domains of the functions.

12. Yes, no, no.

Chapter 5

4. Hint: try $f(x, y) = 1/q$ if y is rational and $x = p/q$ in its lowest terms, $f(x, y) = 0$ otherwise.

13. Hint: what about the boundary?

Index of notation

Index